DATA MANAGEMENT FOR RESEARCHERS

Data Management for Researchers

Organize, Maintain and Share your Data for Research Success

Kristin Briney

RESEARCH SKILLS SERIES

Pelagic Publishing | www.pelagicpublishing.com

Published by Pelagic Publishing
www.pelagicpublishing.com
PO Box 725, Exeter EX1 9QU, UK

Data Management for Researchers

ISBN 978-1-78427-011-7 (Pbk)
ISBN 978-1-78427-012-4 (Hbk)
ISBN 978-1-78427-013-1 (ePub)
ISBN 978-1-78427-014-8 (Mobi)
ISBN 978-1-78427-030-8 (PDF)

This book should be quoted as Briney, K. (2015) *Data Management for Researchers: Organize, Maintain and Share your Data for Research Success*. Exeter: Pelagic Publishing, UK.

British Library Cataloguing in Publication Data
A catalogue record for this book is available from the British Library.

Cover image: Philartphace/Shutterstock.com

Typeset by XL Publishing Services, Exmouth

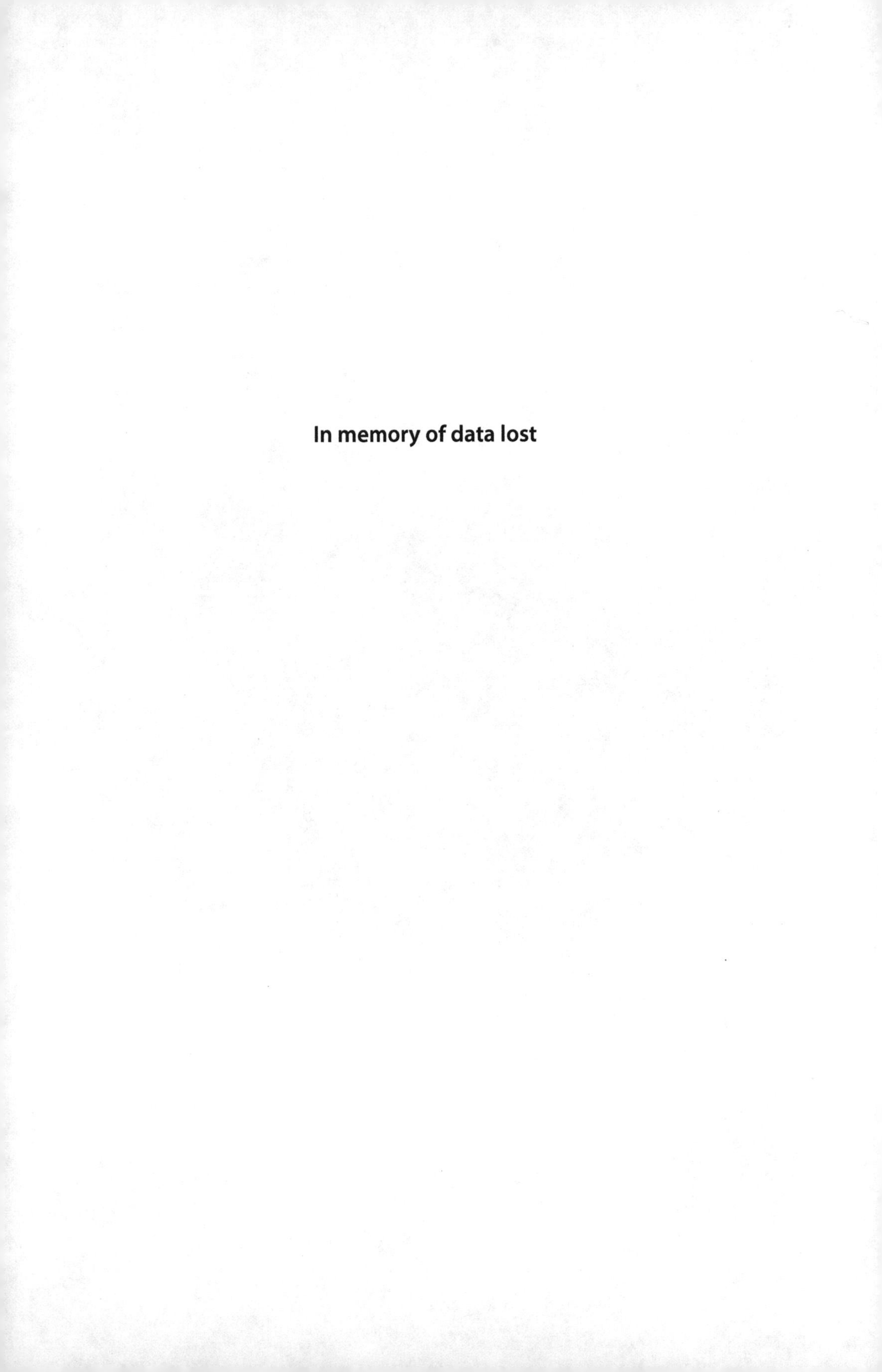

In memory of data lost

CONTENTS

ABOUT THE AUTHOR

Kristin Briney began her career as a research chemist before becoming more interested in (and frustrated by) the management of research data. She currently works in an academic library, advising researchers on data management planning and best practices for dealing with data. Kristin holds a PhD in physical chemistry and a Master's degree in library and information studies, both from the University of Wisconsin-Madison. She blogs about practical data management at www.dataabinitio.com.

ACKNOWLEDGEMENTS

Writing a book is a substantial project that cannot be done without significant support from other people.

I want to thank two people who particularly helped to make this book happen. A special thank you to Dorothea Salo for teaching me data management and encouraging me to write this book, and to Tim Gritten, my supervisor, for giving me time out of my normal duties to research and write.

Thank you also to my reviewers Ariel Andrea, Andrew Johnson, Trisha Adamus, Jeremy Higgins, Dorothea Salo, Abigail Goben, Amanda Whitmire, Margaret Henderson, and Brianna Marshall for reading over chapter drafts and giving me helpful feedback.

I'm also grateful for the work done by the blog Retraction Watch. Without their chronicle of corrections to the scientific record, this book would be much less colorful.

Finally, thank you to my husband who read over drafts, offered a useful researcher perspective, and generally kept the house running while I wrote a book.

1

THE DATA PROBLEM

On July 20, 1969, Neil Armstrong climbed out of his spacecraft and placed his feet on the moon. The landing was broadcast live all over the world and was a significant event in both scientific and human history. Today, we can still watch the grainy video of the moon landing but what we cannot do is watch the original, higher quality footage or examine some of the data from this mission. This is because much of the data from early space exploration is lost forever.

Among the lost data are the original Apollo 11 tapes containing high-quality video footage of the moon landing. Their loss first came to light in 2006 (Macey 2006) and NASA personnel spent the next three years searching for the tapes across multiple continents before concluding that they were likely wiped and reused for data storage sometime in the 1970s (NASA 2009; O'Neal 2009; Pearlman 2009). Other data from this era fared better but at the cost of significant time and money. The Lunar Orbiter Image Recovery Project (LOIRP 2014), for example, spent years and well over a half a million dollars recovering images taken of the moon by the five Lunar Orbiter spacecraft missions preparing for the moon landing in 1969 (Wood 2009; Turi 2014). The project required finding specialized and obsolete hardware to read the original magnetic tapes, reconstructing how to process the raw data into high-quality images, decoding the labeling scheme on each of the tapes, and doing all of this with little to no documentation. Only the cultural importance of the data on these tapes, such as the first image of the earth as seen from the moon, made such efforts worthwhile.

The story of this momentous occasion in scientific history ends with an all-too-common example of failing to plan for data management. Almost 50 years later, researchers are still inadvertently destroying data or having trouble finding data that still exists. A recent study of biology data, for example, found that data disappears at a rate of 17% per year after publishing the results (Vines *et al.* 2014). Another estimate says that 31% of all PC users have suffered complete data loss due to events outside of their control; this correlates with 6% of PCs losing data in any given year (Anon 2014a). Unfortunately, very few of us have significant resources – as with the lunar data projects – to recover our own data when something happens to it. Lost, misplaced, and even difficult to understand data represents a real cost in terms of time and money. Fortunately, there are practices you can use to make it easier to find and use your data when you need it; those practices are collectively called "data management".

At its most essential, data management is about taking care of your data better so that you don't experience small frustrations when actively working with your data, like having trouble finding documentation for a particular dataset, or bigger problems after a project ends, like lost data. Having well-managed data means that you can find a particular dataset, will have all of the notes you need, can prevent a security breach, can easily use a co-worker's data, and can manage the chaos of an ever-growing number of digital files. Basically, many of the little headaches that researchers often encounter around data during the research process can be prevented through good data management. Just as you need to periodically clean your home, so too should you do regular upkeep on your data.

The good news is that dealing with your digital research data does not have to be difficult, though it is different than managing analog content. This book will show you many practices you can use to take care of your research data better. The ultimate goal is for you to be able to easily find and use your data when needed, whether it is historic 50-year-old data or the critical dissertation data you collected last week.

1.1 WHY IS EVERYONE TALKING ABOUT DATA MANAGEMENT?

"Data management" is a relatively new term within research, arising in the mid-2000s with funder requirements for both data management and data sharing. Such mandates gained momentum in the UK with the 2011 Common Principles on Data Policy from Research Councils UK (Research Councils UK 2011) and in the United States with the National Science Foundation's data management plan requirement in 2011 (NSF 2013). Data management and sharing policies are now becoming commonplace in science, with recent adoption by journals such as *Science* (Science/AAAS 2014), *Nature* (Nature Publishing Group 2006), and *PLOS* (Bloom 2013). The overall trend is for increased data management but let's examine why this trend exists in the first place.

We cannot discuss the rise in data management requirements without examining its partner, data sharing. The two concepts often pair together in addressing similar problems in the scientific process, such as limited resources, reproducibility issues, and advancing science at a faster rate (Borgman 2012). The pairing also occurs because well-managed data requires less preparation for sharing. Taken as a whole, most of the reasons why you are now required to manage and share your data are external, though there are many personal benefits to having well managed research data, which we will examine throughout the book.

One of the main reasons behind the implementation of data management and sharing requirements relates to money. The rise of data management requirements roughly coincided with the global economic recession of the late 2000s when many research funding groups faced smaller budgets. With limited resources, funders want to be sure that researchers are making the best use of those resources, for example, by preventing the common occurrence of losing data at the end of a project

(Vines *et al.* 2014). Public funders face additional pressure to make research products like articles and data available to the public who support the research; the current default is that these resources are locked behind paywalls, or are not even made available in the first place. By requiring data management and sharing, funders can not only stem the loss of important data but also provide accountability to those ultimately paying for the research. As an added benefit, any data reuse – either by the original researcher or other researchers – means that the same amount of money will result in more research because data usually costs more to collect than to reuse. Therefore, many research funders see data management and sharing requirements as advantageous.

Another key reason for data management and sharing policies is the prevalence of digital data in scientific research. Research data is digital on a scale never seen before which opens up a whole new set of possibilities in scientific research. First, digital data is shareable in a way not easily done with physical samples and paper-based measurements. It's simple to copy and paste digital values, attach a file to an email, or upload a dataset to the web, meaning it's easy to share research data. We are discussing data sharing so much more because it's actually possible to share data on a global scale. We also generate more data than ever before. The world created an estimated 1.8 zettabytes (1.8 x 10^{21} bytes) of digital content in 2011, a number which is expected to be 50 times bigger in 2020 (EMC 2011; Mearian 2011). Researchers are seeing a similar increase in not only their own data but an added availability of external data. This changes the types of analysis scientists can do. You can now perform meta-analysis or correlate your data with third-party data you would otherwise not be able to collect. Researchers lacking funding or from less developed countries are now able to take part in cutting-edge research because of shared data. Basically, by sharing research data, we open up scientific research to many new types of analysis and can increase scientific research at a faster rate.

In spite of all the benefits of digital data, digital data is fragile. It you do not care for your data, many things can go wrong. Storage devices become corrupt, files are lost, and software becomes out-of-date and media obsolete. Most people have digital files from ten years ago that they cannot use. However, this does not have to be the fate of your research data. Data is a valuable research product that should be treated with care and data management requirements are one way to make that happen.

Finally, data management and sharing policies arose in response to recent reproducibility crises in several scientific disciplines. For example, prominent psychology researcher Deiderick Stapel prompted a reproducibility crisis in his field when it came to light that he committed widespread data fabrication (Bhattacharjee 2013); Stapel amassed over 50 retractions as a result (Oransky 2013a). In economics, a graduate student, Thomas Herndon, proved that the seminal paper supporting economic austerity policies was fundamentally flawed after examining the raw dataset behind the paper (Alexander 2013). In medical research, a study of cancer researchers found that half of the survey respondents had trouble reproducing published results at some point in time (Mobley *et al.* 2013). Clearly, there is a reproducibility crisis in scientific research as these stories represent just a few highlights of reproducibility issues in recent years. Adding to this is the fact that it can be difficult to tell from an article alone whether a study is reproducible because "a scientific publication is not the scholarship itself, it is merely advertising of the scholarship" (Buckheit and Donoho 1995). The creation of data management and sharing policies is one response to this reproducibility crisis, as these policies help ensure that data is available for review should questions arise about the research. Misconduct investigations are also starting to look at data management. For example, investigations leading to the high-profile retractions of two STAP (stimulus-triggered acquisition of pluripotency) stem cell papers in 2014 "found inadequacies in data management, record-keeping and oversight" (Anon 2014b). With a growing number of retractions in recent years (Fanelli 2013; Steen *et al.* 2013), good data management increasingly needs to be part of a good defense against charges of fabrication or even more political attacks on high-profile research.

All of these issues – limited funding, the ease of sharing digital data, the availability of new types of analysis, the fragility of digital data, and reproducibility issues within scientific research – coincided to provide an optimal environment for the creation of data management and sharing policies. Along with new policy requirements, they will continue to fuel the drive toward better data management in scientific research.

1.2 WHAT IS DATA MANAGEMENT?

While many researchers were introduced to the concept of data management through a funder's requirement to write a data management plan, there's actually a lot more to data management than planning. Moreover, it's the data management you do after writing the plan that really helps in your research. This section covers what "data management" actually entails, but first we need to define what is meant by the "data" portion of "data management".

1.2.1 Defining data

Defining research data is challenging because data by its very nature is heterogeneous. Research fields are diverse and even specific subfields use a huge variety of data types. So instead of limiting ourselves to one definition of data – which likely doesn't cover everything – let's explore several definitions.

Data – singular or plural?

One way to prompt an argument among data specialists is to ask if the word "data" is singular or plural. Should it be "data is" or "data are"? While some people use whichever is more comfortable for them, others have very strong opinions about which is correct. Until recently, I was on the data-are-plural side of this skirmish but decided, in the course of writing this book, to use the data-is-singular viewpoint. Here are the reasons why I ultimately think that a singular "data" is preferable.

The major reason for using "data" as a singular noun is that many people stumble over reading "data" as a plural. I had several people comment that, while they understood the reasoning behind making "data" plural, it was simply too distracting. The meaning of the text gets lost while the brain figures out how to process the plural form of "data". Using "data" as a singular noun simply makes for an easier read with a clearer message.

The other reasoning behind using "data" as a singular noun comes from Norman Gray (Gray 2012). Gray wrote an essay on why "data" is a singular noun that includes both the history of the word "data" and its grammatical considerations. The main theme is that, while the word "data" originally derives from Latin, how the word is used in English ultimately matters more than how the word's origins suggest it should be used. This means that the commonly used singular form of "data" is correct even if the rules of Latin suggest that "data" should be plural. Additionally, since we hardly ever use the Latin singular version of "data" – datum – the word "data" is the de facto singular version in the English language. Just like "agendum" has given way to "agenda", so too do we use "data" as an all-encompassing singular noun.

This book, therefore, uses "data" as a singular noun and I hope you find it easier to read.

In the United States, research data created under federal funding falls under the definition of data in OMB Circular A-81:

> Research data means the recorded factual material commonly accepted in the scientific community as necessary to validate research findings, but not any of the following: preliminary analyses, drafts of scientific papers, plans for future research, peer reviews, or communications with colleagues. This "recorded" material excludes physical objects (e.g., laboratory samples). Research data also do not include:
>
> (i) Trade secrets, commercial information, materials necessary to be held confidential by a researcher until they are published, or similar information which is protected under law; and
>
> (ii) Personnel and medical information and similar information the disclosure of which would constitute a clearly unwarranted invasion of personal privacy, such as information that could be used to identify a particular person in a research study. (White House Office of Management and Budget 2013)

This definition is very broad, covering anything necessary to validate research funding, but is helpful in that it outlines what definitely is not research data. These exclusions are particularly useful in complying with data sharing requirements to know what you are not required to share.

More globally, the Organisation for Economic Co-operation and Development (OECD), consisting of 34 member nations, provides a similar definition in their "Principles and Guidelines for Access to Research Data from Public Funding":

> "Research data" are defined as factual records (numerical scores, textual records, images, and sounds) used as primary sources for scientific research, and that are commonly accepted in the scientific community as necessary to validate research findings. A research data set constitutes a systematic, partial representation of the subject being investigated.
>
> This term does not cover the following: laboratory notebooks, preliminary analyses, and drafts of scientific papers, plans for future research, peer reviews, or personal communication with colleagues or physical objects (e.g. laboratory samples, strains of bacteria and test animals such as mice). (Organisation for Economic Co-operation and Development 2007)

This report focuses on the sharing of digital datasets so this definition of data skews toward digital content. In actuality, physical samples can be research data and, in some cases, fall under data sharing requirements (see Chapter 10).

You may also see data defined by type. Just as social science data often falls into one of two categories – quantitative or qualitative – so too does scientific data fall into specific groups. For scientific research data, those categories are:

- Observational data
- Experimental data
- Simulation data
- Compiled data

Observational data results from monitoring events, often at a specific time and place, and yields data such as species counts and weather measurements. Scientists produce experimental data in highly controlled environments so that similar conditions will always result in similar data; examples of experimental data are spectra of chemical reaction products and the measurements coming from the Large Hadron Collider. The third category of scientific data is simulation data, which results from computer models of scientific systems. Global warming simulations and optimized protein folding pathways represent two types of simulation data. The final category, compiled data, applies when you amass data from other sources for secondary use, such as performing meta-analysis or building a database containing a variety of data on one topic. While not perfect, most of the content we consider to be scientific research data fits into one of these four categories.

For this book, we'll use a broad definition of research data: data is anything you perform analysis upon. This means that data can be spreadsheets of numbers, images and video, text, or another type of content necessary for your research. Data can also be physical samples or paper-based measurements, though analog content usually has fairly established management practices. In the end, it's simply too much to try to define every possible type of data and a broad definition of data allows you, the researcher, to be generous in identifying content that you need to manage better.

1.2.2 Defining data management

If you've ever gotten halfway through a project and thought "why didn't I write down that information?" or "where did I put that file?" or "why didn't I back up my data?" then you could benefit from data management. Data management is the compilation of many small practices that make your data easier to find, easier to understand, less likely to be lost, and more likely to be usable during a project or ten years later. Data management is fundamentally about taking care of one of the most important things you create during the research process: your data.

Data management involves many practices. We will examine these more in Chapter 2. Briefly, data management includes data management planning, documenting your data, organizing your data, improving analysis procedures, securing sensitive data properly, having adequate storage and backups during a project, taking care of your data after a project, sharing data effectively, and finding data for reuse in a new project. Such a wide range of practices means that data management is something you do before the start of a research project, during the project, and after the project's completion.

The reason you do all of these data management practices is so that you don't get stuck without your data when you need it or end up spending hours trying to reconstruct your data and analysis. The rule of thumb is that every minute you spend managing your data can save you ten minutes of headache later. Dealing with your data can be a very frustrating part of doing research but good data management prevents such tribulations.

Data management vs data management plans

If you are coming to data management because you need to write a data management plan (DMP), it is important to know that writing a DMP and actually managing your data are two different things. A data management plan describes what you will do with the data during and after a project while data management encompasses the actions you take to care for your data. Each requires its own skill to execute but understanding data management makes it much easier to write a data management plan. So while a DMP is a helpful outline (and likely satisfies funder requirements), data management is the thing that will actually make it easier for you to use your research data during and after a project.

This book is about data management. More specifically, this book is about all of the practices you can use to make your data easier to find and understand, more secure, and more likely to be usable for ten years into the future. My goal is to create a practical guide to help you deal with your research data.

1.3 WHY SHOULD YOU DO DATA MANAGEMENT?

My personal interest in data management began out of frustration. I was a PhD student in Physical Chemistry at the time and was studying the effect of vibrational energy on photoswitching kinetics in solution using ultrafast laser spectroscopy. Like many graduate students, I lacked the proper knowledge and tools for managing the data generated during my experiments and analysis. It was difficult enough to deal with my own data but what really frustrated me was the part of my thesis project based on studies conducted by a previous graduate student. When questions arose about how her analysis related to my own, I ran into a full data management nightmare. Basically, I either had to rebuild her analysis from her notes and old files or repeat the experiment to recreate her data and analysis. When I failed at the former I ended up doing the latter, which – despite being a terrible waste of time and resources – was the only way for me to fully understand the previous study. The whole experience left me with a distinct impression that there must be a better way to manage data.

Unfortunately, my data management frustrations are not unusual. Since becoming a data management specialist, I've talked to many researchers struggling to use someone else's data, coping with disorganized files, and dealing with lost data, among other issues. The root of the problem is that most of our research data is now digital but we are not yet accustomed to managing digital content. You can put a lab notebook on a shelf and still use it in ten years but you cannot necessarily do this with digital data. The good news is that there is a better way to deal with your digital content and that is through conscious data management.

I firmly believe that, even if you purchased this book because you need to know how to write a data management plan for a funding agency, data management can actually make a difference in your research. All it requires is a little conscious forethought, making small practices routine, and thinking about your data as an important research product. Take a little time to care for your data well and your data will work for you, not against you.

I want to end by saying that data management does not have to be difficult. Good data management is just the cumulation of many small practices that you make a routine part of your research. And any little bit you do to improve your data management helps. For example, if all you get from this book is that you should take better notes (see Chapter 4) or have an extra offsite backup of your data (see Chapter 8), then your data is easier to work with and safer than before. So, start small with data management and take incremental steps until you have well-managed data. You don't have to take every recommendation in this book or do everything all at once, but I challenge you to adopt at least one practice from this book to improve your data management. Make your data work better for you by taking care of it through good data management.

2

THE DATA LIFECYCLE

In 1665, the Royal Society in London created the very first scientific journal, *Philosophical Transactions of the Royal Society* (Royal Society Publishing 2014). This changed the way that scientists communicated, from writing to each other by letter to publishing research articles for a wider audience. While publication has since become the standard, it was not universally accepted at the time. Some scientists felt that their research should not be shared with anyone except those scientists of which they specifically approved. Over time, however, publishing research results as an article became the accepted way to get credit for research.

The transition from private communication to public publishing echoes the recent emphasis on sharing and preserving research data. As with research results prior to 1665, researchers have traditionally only shared data privately with select peers. This is changing, however, as research funders are starting to require data sharing so that researchers now publish both articles and data. The idea of data sharing is not without controversy, which, if we draw from the example of the first research journal, means that the process of research is fundamentally changing.

Just as publishing results in a scientific journal changed the way we do research, so too will data preservation and sharing. The use of data beyond the original project, while already occurring intermittently in research, means a significant change in the way we view data throughout the research process. Data becomes an important research product in its own right. This also entails a change in how we manage data so that we can actually preserve, share, and reuse data beyond the project's end.

2.1 THE DATA LIFECYCLE

To understand data's role in the overall research process, and thus how to manage data better, we must start by breaking the research process down into the steps that make it up. The use of this so-called "data lifecycle" is common within data management as it helps identify the role data plays at different points in a research project. For example, data is central to the data acquisition and analysis portion of the research process, meaning that it will likely need lots of active management during this stage. By breaking data use up into a lifecycle, we are better able to understand the place that data management takes in each part of research.

Therefore, in order to understand the different strategies needed to manage data over time, it is useful to first examine the role of data in each part of the research process. And since data's role in research is currently changing, we will examine the old data lifecycle and the new data lifecycle.

2.1.1 The old data lifecycle

The current data lifecycle has been in existence since the publication of research articles became the standard almost 400 years ago. The cycle starts with project planning, continues with data acquisition and analysis, and concludes with the publication of research results (see Figure 2.1). This is a simple view of the research process that helps us frame data's role within research.

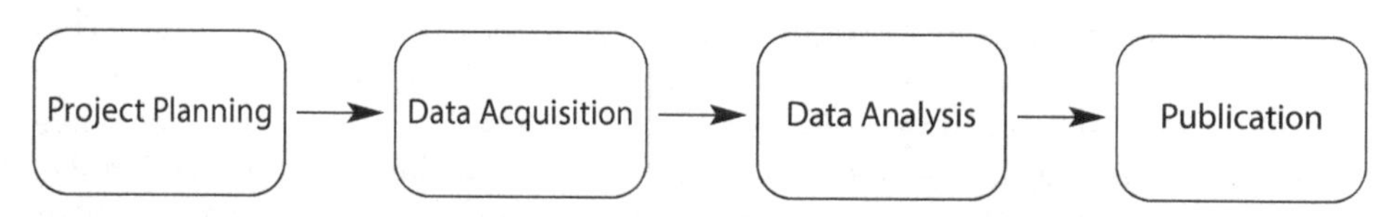

Figure 2.1 The old data lifecycle

Data occupies an important place in the middle of this process, with acquisition and analysis being very data-centric activities. However, data also plays a small part in the other stages of project planning and publication – for example, you must think about what data you want to acquire during project planning, and you represent data as figures and tables in the publication process. So data is important in all stages of the current research process but most critical to acquisition and analysis.

Taken as a whole, this shows a picture of data as a means to an end: article publication. This cycle does not reward the use of data for much beyond analysis, so data is a by-product of research instead of an important research product, like articles. One of the biggest indicators that data is not viewed as a research product is that data is usually lost after the end of a study. A study of biology research data conducted by Vines *et al.* in 2014 found that the availability of data after publication falls by 17% per year (Vines *et al.* 2014). So the older the article, the less likely it is that the data can be found and used. This is a problem because research data often has value beyond the original study, both within the original laboratory and outside of it. Yet because the current research process does not value data, there has been little incentive to keep data after publication.

This system worked well for hundreds of years but the prevalence of digital data in research means that we can do more with research data beyond losing it at the end of a project. Between data preservation, sharing, and reuse, data can now be a research output in its own right. This new emphasis on data means that we need an updated data lifecycle.

2.1.2 The new data lifecycle

The new data lifecycle adds data sharing, data preservation, and data reuse as steps in the research process. As a whole, the cycle includes: project and data management planning, data acquisition, data analysis, article publication and data sharing, data preservation, and data reuse (see Figure 2.2). This lifecycle assigns a greater importance to research data than the previous cycle by making it an actual product of research.

The new lifecycle is also a true cycle, in that data from a previous project can feed into a new project and cause the cycle to begin over again. Data does not default to being lost at the end of a project and instead is preserved and reused. Additionally, it does not have to be your own data feeding into the next project; the increased prevalence of data sharing means that more outside data is available for you to use and correlate with your own research.

Looking into the importance of data in this new cycle, we see that data is a significant part of every stage of research. Where project planning and article publication nominally involved data before, we now have data management planning and data sharing making data an integral part of research at these two stages. Taken as a whole, this new research data lifecycle is deeply rooted in data and presents data as a useful research product.

2.2 THE DATA ROADMAP

I like using the data lifecycle as an outline for data management because data management techniques slot nicely into the different parts of the data lifecycle. This is not only helpful for thinking conceptually about data management but also for managing data during a project – you principally have to worry about the data management practices for whatever part of the lifecycle you are currently in. For these reasons, I use the data lifecycle as the rough layout of this book.

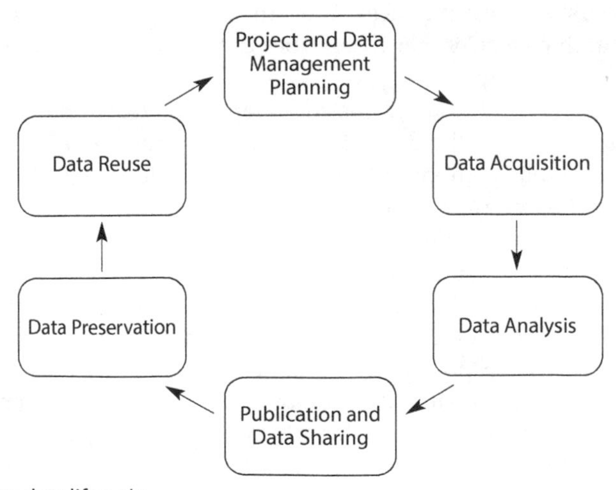

Figure 2.2 The new data lifecycle

The correlation between the different stages of the new data lifecycle and the data management strategies that they entail is what I'm calling the "data roadmap" (see Figure 2.3). As you work through a research project and arrive at different steps in the lifecycle, the data roadmap highlights the actual data management practices that will help you at that particular stage of the project. The goal is that by the time you reach your destination, the end of the project, you will have well-managed data that you can easily reuse and that will live well beyond the end of the project.

2.2.1 Following the data roadmap

To start your journey on the data roadmap, start at the beginning of the data lifecycle with project and data management planning. In practice, this means strategizing for how you will manage your data, creating a data management plan, and being aware of the data policies that apply to your work. This is in addition to the general project planning and literature reading you normally do at the start of a project. Refer to Chapter 3 for more information on data management planning.

The next stop on the roadmap is data acquisition. In terms of data management, this means two groups of techniques: documentation and data organization. Documentation can include anything from using lab notebooks, metadata, and protocols, as well as other useful documentation formats such as README.txt files and data dictionaries. Chapter 4 covers these documentation structures in more detail. Organization strategies for data include file organization, file naming, how to document conventions, and using a database. Chapter 5 goes into further specifics on each of these practices. You should use both documentation and organization practices while you are actively collecting data.

The next destination on the map corresponds to the analysis portion of the data lifecycle. Data management plays less of a role during data analysis but there are still some useful considerations. These include understanding the difference between managing raw and analyzed data, as well as performing data quality control and following spreadsheet best practices. Data management strategies also extend to managing research code. Chapter 6 describes data management strategies relevant to this stage of the journey.

We'll now take a little detour on the roadmap to consider data storage, which is relevant to many stages of the data lifecycle. Data storage and backup, covered in Chapter 8, is particularly important during the lifecycle stages of data acquisition and analysis. The lifecycle stages of data sharing and data preservation require long-term storage. Chapter 9 examines what data to retain and how long to retain it, how to prepare data for the long term, and outsourcing data preservation to a data repository. If you have sensitive data, such as personally identifiable information, you'll want to plan your journey even more carefully. Consider what types of data require extra security, how to keep data secure, and how to anonymize data. Chapter 7 discusses each of these topics in greater detail. Continuing with the analogy, storage is like the car that carries your data along the data roadmap, so you want to make sure it's sufficient for the journey.

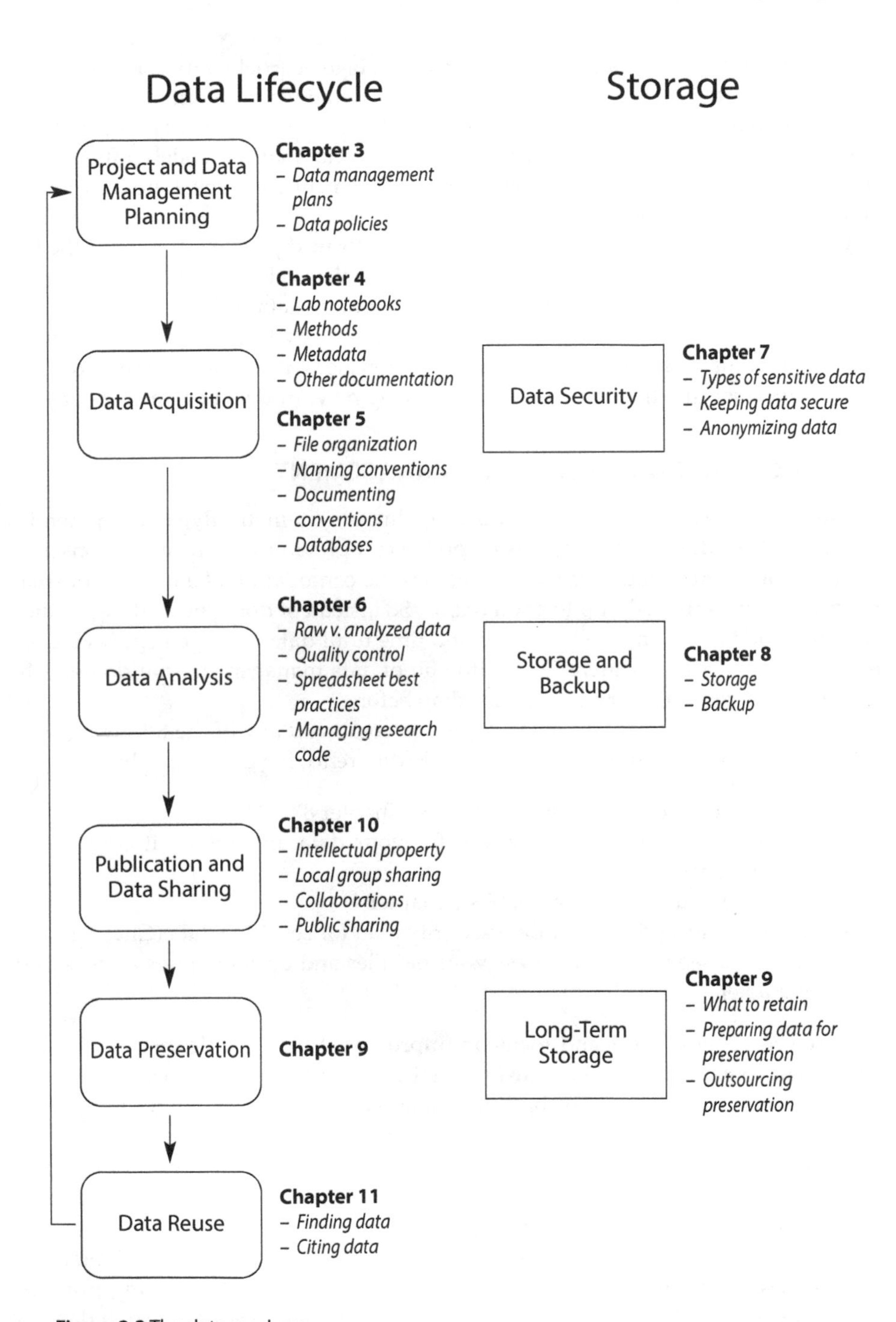

Figure 2.3 The data roadmap

We reach our final destination with the lifecycle stage of article publication and data sharing. For data management in this stage, you should worry about the legal framework for sharing, which includes copyright, licenses and contracts, and patents. Data sharing also exists in different regimes – local sharing, collaborations, and public sharing – each with their own data management concerns. Refer to Chapter 10 for more information on these topics.

Once we reach our destination and consider our next journey, we are in the last stage of the data lifecycle (and the one that begins the cycle over again): data reuse. Data management concerns at this stage include finding data and citing data. Chapter 11 covers these topics further. From here, you are ready to start a new journey on the data roadmap. Hopefully your next excursion along the data roadmap proceeds even more smoothly than the previous as managing your data becomes easier.

2.3 WHERE TO START WITH DATA MANAGEMENT

While you can work through the data roadmap systematically, I recommend a slow, selective adoption of practices if you haven't previously done much conscious data management. Good data management is the compilation of a number of small routine practices that add up to good habits. So instead of doing everything at once, start by working on one new strategy at a time until data management becomes a routine part of your research. Every little bit of data management you do helps by making your data safer or easier to use than before.

If you are unsure about where to begin, I recommend any of the following as an achievable starting point that can make a big difference in your research:

- Make sure you have reliable backups (Chapter 8)
- Develop an organization system for your data and follow it consistently (Chapter 5)
- Improve your note-taking habits (Chapter 4)
- Review and update your data security plan for sensitive data (Chapter 7)
- Check to see if you can access your old files and update things as necessary (Chapter 9)

Choose one of these tasks and focus on improving it for a whole month. For note taking and organization, consciously practice doing things the correct way until these practices are routine. For backups, data security plans, and old files, consult your co-workers to see what systems they use and if you can find a common system to use together. Basically, do a little work on that task over the course of the month until you are happy and comfortable with your new data management practice. And then choose a different topic for the subsequent month.

The data lifecycle and the data roadmap are useful structures for thinking about data management but, in practice, data management is about following practices that make it easier for you to organize, maintain, and share your research data. So just as you do not have to slavishly follow the data roadmap, you also do not have to adopt every data practice in this book. I lay out a range of practices to show the ideal, but ultimately you should find a system that makes your data easier to use, protects your data from loss, and generally fits well into your research workflow. Manage your data so that your data works better for you.

2.4 CHAPTER SUMMARY

Data plays many important roles throughout the research process and these roles are framed by the data lifecycle. Traditionally, the data lifecycle emphasized data mostly for analysis, but new focus on data management and sharing has expanded both the lifecycle and data's roles within it. The new lifecycle is also a true cycle, with old data feeding into new research projects.

This book uses the data lifecycle as a roadmap for tackling data management by matching data management practices with particular portions of the new data lifecycle. This should help you prioritize which data management strategies to address at different stages of a research project.

While you can work through the whole data roadmap sequentially, consider starting with one of these practices: creating reliable backups, organizing your data, improving your note taking, reviewing your data security plan, or surveying your old files. It also helps to make data management routine; try focusing on one strategy per month before moving onto the next. Adopting even a few data management practices can make it easier to use and reuse your data.

3

PLANNING FOR DATA MANAGEMENT

Consider this approach to data management:

> I will store all data on at least one, and possibly up to 50, hard drives in my lab. The directory structure will be custom, not self-explanatory, and in no way documented or described. Students working with the data will be encouraged to make their own copies and modify them as they please, in order to ensure that no one can ever figure out what the actual real raw data is.
>
> Backups will rarely, if ever, be done. (Brown 2010)

C. Titus Brown's satirical data management plan highlights the usual fate of data without a data management plan. Data lacks order and security, making it in no way useful for anyone in the lab (and often not even for yourself). If this sounds like your current data management strategy, it's time to come up with a new plan for managing your data.

The most import time to address the management of your research data is before you even start to collect that data. Planning ahead means that you can properly organize, document, and care for your data throughout the project and afterward. The alternative is realizing toward the end of a project that you are missing data, missing notes, or do not have the proper permissions to analyze and publish a particular dataset. Indeed, a little data planning up front can save you a load of time and trouble later in the project.

There are many aspects to planning for proper data management, which is the focus of this chapter. First, you should know what to plan for. Second, you should know how to create a data management plan, either for your own purposes or because you are required to create a plan. Finally, data management plans exist within a framework of various data policies, which I cover at the end of the chapter.

3.1 HOW TO PLAN FOR DATA MANAGEMENT

Before you directly work with your research data, it helps to think about all of the data management strategies you want to utilize during and after your research project. There are many choices to make, though some have greater consequences than others. Additionally, the longer the duration of time you need to keep the data and the more people who need to use it, the more important it is to plan ahead how to manage everything. I understand that you may choose not to address all of the recommendations in this book, but it is better to make an informed decision to not take up a practice rather than be unaware of the fact that you need it.

3.1.1 The importance of planning for data management

As you start planning for data management, you should know that good data management takes time and effort. Data, like all other things in this universe, follows the Second Law of Thermodynamics and will tend toward chaos if not occasionally provided with energy. In practice, this means putting thought into customizing the best solutions for you and your data and directing energy into following through with this plan. Generally, the more effort you put into data management upfront, the easier it becomes to work with your data later. The old adage "an ounce of prevention is worth a pound of cure" holds true even for data.

Strategies for data management can be as varied as data itself. There are some commonalities, but the actual execution is up to you. Therefore, I will make a number of recommendations in this book, such as file naming conventions, that lay out the basic principles and allow for customization on your part. The ultimate goal is for you to create a system that works for you and your co-workers, realizing that you are the best judge on how to make that happen.

All of this is to say that if you want the most from your data, you must spend time thinking about how to care for it. If you come up with the best system for your data before you start a project, you set yourself up for success in managing your data during the project and afterwards.

3.1.2 How to customize data management to your needs

Much of your effort in planning for data management will be spent finding the systems that work best in your research. Good data management is a balance between best practices and your individual needs, and part of achieving that balance involves understanding those needs. Begin by asking yourself the following questions:

- What types of data do I have? How much do I have?
- Do I use any third-party data?
- What data tools and technology are readily available to me?
- How much do I collaborate? Is this internal or external to my institution?
- How long must I keep my data?
- Will I share my data?
- Does my data have security concerns, such as personally identifiable information?
- What does my funder/institution/employer require?
- Is there anything particular in my research workflow that might affect my data management?
- What problems with my data do I often encounter?

The answers to these questions will have a direct impact on the strategies you use to customize your data management plan.

Using this profile of your data, examine the topics outlined in the data roadmap from Chapter 2 (see Figure 3.1) to determine which are most important to the management of your data. Perhaps your local institution provides all of the necessary storage and backup systems but you need an organization scheme that will work for you and your collaborators. In that case, you would prioritize the categories of data organization and collaborations over data storage.

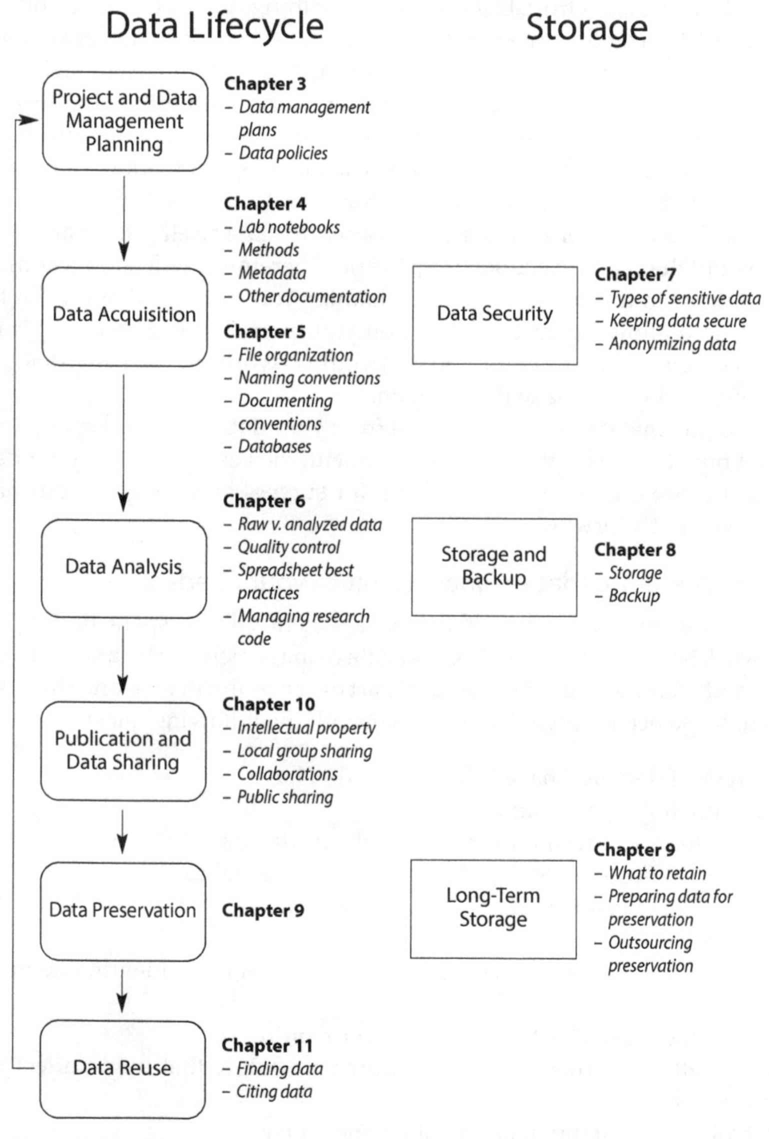

Figure 3.1 The data roadmap

Once you lay out your general priorities, examine each topic in greater detail. Consider different data management approaches and weigh them against your data, your available tools and your research workflow; refer to subsequent chapters in this book for data management strategies you can adopt. Also consider how you might customize different approaches to suit your needs. For example, say you share a large amount of data with your collaborators and you need a documentation system that allows for active collaboration, enables easy searching, and does not significantly alter your current workflow for taking research notes. Depending on your exact needs, an electronic laboratory notebook might be the best solution, while in other cases, the adoption of a formal metadata schema and a shared database may be preferable. The appropriate data management solution must always work for you and your specific data.

Remember that good data management need not be difficult or complex, but instead is often the summation of many small practices over a range of data-related topics. The best solutions are the ones that become a routine part of your research workflow. For example, consistently using a file naming system can mean the difference between order and chaos in your data files. Certainly, good data management requires effort, but the more practices you can make automatic the better chance you have at successfully caring for your data.

Your ultimate goal is to create a system that is comprehensive, yet achievable. This point comes at the intersection of best practices, your data, and your research workflow. Do a little research, be a little creative, and come up with a list of strategies to adopt. From here, the next step is to document and elaborate on your data management solutions. The subsequent section covers the creation of a written data management plan and identifies many small details to include in your plan.

3.2 CREATING A DATA MANAGEMENT PLAN

You should create a data management plan regardless of whether or not one is required of you. While funder requirements have introduced many researchers to the concept of a data management plan, these plans are very useful beyond such mandates. At its core, a data management plan is a document that lays out how you will deal with your data. As previously mentioned, the best way to improve your data management practices is by planning ahead and the data management plan is the thing that actually codifies what you will do. For that reason, I recommend every scientist or group create a data management plan before starting a project.

3.2.1 Why create a written data management plan?

Funding agencies require data management plans because they understand the value and fragility of research data. Without proper management and care, important data often does not exist much past the completion of the grant. Indeed, a 2014 study found that after article publication, the availability of the published data falls off by 17% per year (Vines *et al.* 2014). As funders invest significant resources into supporting research, the data management plan is one way to ensure that these resources do not go to waste.

Even if one is not required, a written data management plan is an incredibly valuable tool for you, the researcher. First, you have a document to refer to when there is a question on how to manage your data. Second, a written plan is especially useful when you are actively sharing data and other research materials with collaborators. Having everyone use the same conventions for naming, organization, storage, etc. makes it decidedly easier to use someone else's data. Even without active data sharing, research groups may wish to adopt a group data management plan. An overarching plan makes it easier for anyone in the group, and especially the group leader, to find and use a person's data, even after a member leaves the group. Such a plan also helps introduce new group members to data management because the plan lays out the expected management procedures. Finally, it's just good practice to plan for data management as it saves you time and effort later.

3.2.2 What a data management plan covers

The specific content of your data management plan will depend on your data, your needs, and any external requirements for your plan, but generally a data management plan addresses the following major topics:

1. What data will you create?
2. How will you document and organize your data?
3. How will you store your data and, if necessary, keep it secure?
4. How will you manage your data after the completion of the project?
5. How will you make your data available for reuse, as necessary?

The answers to these questions need not be complex, but instead should reflect how you want to ideally yet realistically manage your data.

Let's go through each of the five major topics to understand its contribution to the data management plan. I will break down each topic into a range of issues to document in your plan, but realize that you are not required to address every bullet point. Rather, the suggestions show you the scope of issues to consider when planning for data management.

I will add a special note here to say that, as you address each of the major topics, be sure to document any resources devoted to your plan, such as work time and necessary equipment. If a portion of your plan needs special or continued attention, such as overseeing compliance with privacy laws, it's a good idea to designate someone responsible and record that in your plan. This helps establish clear responsibilities and hold people accountable. Additionally, if you are creating a data management plan for a grant application, you should clearly identify any resources you need in order to carry out your plan. This can help you secure extra funding to manage your data properly.

The first topic addressed in your plan should be a general description of your data, including such things as:

- How big will your data be?
- How fast will your data grow?
- What are the likely file formats for the data?
- How unique is the data?
- What is the source of your data?
- Who owns the data?

Basically, you want to provide a general description of your data. While many of these points are not about the actual management of the data, they are critical inputs in determining the best management strategy. For example, a unique dataset should be managed with more care than an easily reproduced dataset, and proprietary file formats require different long-term management to ensure continued access. Therefore, it is good to provide a general description of your data so that you can refer back to it later when you describe the best way to manage it.

Once you describe the data, you should explain how you will organize (see Chapter 5) and document (see Chapter 4) it. This is the second major portion of your plan. Consider:

- How you will document your data?
- What metadata schemas you will employ, if any?
- Will you use supporting documentation such as protocols, data dictionaries, etc.?
- What tools do you need for documentation?
- How you will organize your data?
- Will you use any file naming schemes?
- How you will manage and preserve your documentation?

Many funders prefer you use field-specific metadata schemas and other standards for documentation, to help with interoperability and sharing. It is worth doing a little research into standards for your particular field, not only to make your funder happy but because such standards can be incredibly useful tools for research (see Chapter 4 for more information on structured metadata schemas and standards). Even if you do not use a common standard, you should note any documentation conventions you and your co-workers follow. Likewise for conventions used to organize your data. You can be more or less descriptive, depending on the available space, but it is always best to have a record of your conventions.

The third topic is a very important part of your data management plan and centers on how you will store and back up your data (see Chapter 8). You should note:

- How will you store the data?
- How frequently will you back up the data?
- Who manages the storage and backups?
- What money and resources do you need for storage?

This information is even more important if you have data requiring extra security (see Chapter 7). In this case, you will want to answer additional questions, such as:

- What policies apply to your sensitive data (national and local) and how will you follow any policy requirements?
- What security measures will you put into place?
- Who is allowed to access your data?
- How will you train others on security protocols?
- Who will oversee the management of the security systems?
- Who are the local security resources you consulted/can consult with?
- What money and resources do you need for security systems and training?

Data is sensitive for many reasons, ranging from privacy for human subjects to confidentiality for corporate partnerships or intellectual property claims. No matter the reason, having a written plan makes security and storage expectations transparent for everyone handling the data.

The fourth major data management plan topic builds on the previous by describing what will happen to your data after the completion of the project (see Chapter 9). This is partly a storage question and partly a management one; be aware that practices in both areas change once your data is no longer in active use. The issues to address in this section are:

- What storage and backup systems you will use for your older data?
- For what time duration will you actively maintain your data after the completion of the project?
- What are your plans to migrate storage media over time?
- How will you prepare the data for long-term storage and archiving?
- Who will manage the data in the long term?
- Will you use any third parties to preserve your data?
- What happens to the data if you leave your current institution?
- What money and resources do you need for data preservation?

The issue of data ownership is central to answering many of these questions. For example, the fate of a dataset owned by the person who collected the data, the higher-level individual who directed the research, or the institution that provided resources (physical or monetary) for the work, is often very different. This is because it is up to the data owner to designate how the data will be managed and who will manage it. Either way, realize it takes effort to ensure that files do not get corrupted or lost in the long term and that making a plan is the best first step for preventing data loss.

The fifth and last topic covered in any data management plan concerns data that will be publicly shared, either for policy or personal reasons (see Chapter 10). Even if you do not fall into these groups, it is useful to address these issues from the perspective of sharing your data with your co-workers or future self. Either way, you should address the following questions:

- What data will be shared and what form will it take?
- Who is the audience for your data?

- When and where will you share the data?
- Are there any requirements of the chosen data sharing system?
- How will the data be prepared for sharing?
- Who will be responsible for making the data available?
- Who will be the contact person for questions about the data?
- What money and resources do you need to enable sharing?

My biggest piece of advice here is to use an existing data sharing system. Do not reinvent the wheel! Leveraging an existing system not only saves you time but usually makes your data more permanent and findable than inventing a hosting system for yourself. You want to spend your time worrying about your next research project, so let someone else worry about making your data available.

Finally, it is important to note that, while the five major topics outlined at the beginning of this section form the general foundation for a data management plan, you will likely have specific requirements when creating a plan for a funding application, etc. Such requirements always supersede the five general topics, though these topics still provide a useful narrative framework for your plan.

3.2.3 Creating a data management plan for your research

Think of your data management plan as the map you will follow along the data's journey, helping you navigate the many events you will encounter along the way. Like a map, your plan should be explicit and concise, to the point where someone else can read the plan and immediately understand what you are doing with your data. This is particularly important for research groups, where multiple people will be following the same plan.

If you are based in the United States or the United Kingdom, there are two tools available to help with the creation of a data management plan for a grant application: the DMPTool (University of California Curation Center 2014) and DMPonline (Digital Curation Centre 2014a). The DMPTool is US-based and therefore focuses primarily on US funding agencies. DMPonline is similar but for UK-based researchers and funders. Both tools provide templates for many funding groups that require data management plans. If you are a university researcher in either country, it is well worth using one of these tools to create your data management plan as they lay out all of the requirements and walk you through the creation of a plan.

Realize that the plan you write for a grant application does not have to be the master plan for your research data, especially because funders often put length constraints on data management plans. I highly encourage you to use your grant's data management plan as a draft and expand those areas which need more detail in a master plan. Not only does this save time but you can also plan for as much management as truly benefits your research. The value of data management planning to you and your co-workers should not be dictated by your funding application alone. Make a master data management plan.

For either your master plan or funder plan, your data management plan should be more narrative than a rehash of existing policies. While it's good to quote policy, it is more important to state the exact way in which you will follow the policy. For example, if a privacy policy covers your data, you should state the security systems, de-identification procedures, etc. that put you into compliance with that privacy policy. There are many policies that might apply to your data, and Section 3.3 covers a range of them.

Customizing a data management plan to fit your research takes effort, though this upfront work makes it easier to deal with your data in the long term. It is one thing to know the topics a plan should cover and another to know how to customize those topics to the best benefit of your data. This problem encompasses the entirety of this book and I hope you find many ideas in later chapters that you can apply to your everyday practices and your data management planning.

Finally, periodically review your data management plan as you work on a project to be sure that you are following the plan and to identify anything that needs updating. Even good data management plans can require modification once you establish a good workflow for the project. As the main purpose of having a data management plan is to use it to manage your data, do not be afraid to revise your plan to include new strategies that will make it easier to manage and use your data.

3.3 DATA POLICIES

Much of data management planning is local, but it is important to consider how larger policies affect your plans. Such policies concern a range of topics, come from a number of places, and affect everything from how you store your data to how you share it (University of Cambridge 2010; Goben and Salo 2013; Digital Curation Centre 2014b). You should be aware of all applicable policies so that you can plan and manage your data properly.

3.3.1 Types of policies and where to find them

There are likely to be several policies that apply to your data, though the challenge is that you may not always be aware that they exist. This is because data policies come from a variety of sources, including:

- National legislation
- National policies
- Funder policies
- Regional and local policies
- Institutional policies
- Field-specific community norms
- Journal policies

While you may be aware of the national laws that apply to your data, particularly if you work with human subject information, not everyone is aware of more local policies. Uncertainty about local data policy is especially prevalent in academia where data practices are not as regulated as in many industrial settings.

The best advice I can give you is to determine the policy categories that might apply to your data and do a little research. In general, data policies fall into the following categories:

- Data privacy
- Data retention
- Data ownership
- Data and copyright
- Data management
- Data sharing

The next several sections discuss what these policies cover and provide a starting point for finding the particular policies that apply to your data.

3.3.2 Data privacy policies

The purpose of data privacy policies is to protect sensitive and personal data. Generally, this includes medical information and any information that identifies someone personally, but can also include data that would cause harm if released to the public. Privacy policies do not universally cover all types of sensitive data, but will identify certain types of data as sensitive and place restrictions on access to and storage of that data.

Table 3.1 National laws and international directives on data privacy

Law/Convention	Domain	Description
HIPAA	USA	Regulates acceptable use and storage for personal health information
FERPA	USA	Limits access to student educational records
Data Protection Act 1998	UK	Regulates the storage and use of personal data
Data Protection Directive	European Union	Limits the processing of personal data and directs the creation of similar laws by member nations

The main source for privacy policies is national laws and Table 3.1 highlights several such policies. While all these laws relate to privacy, some are narrower while others are broader. For example, United States regulation on data privacy comes from several narrowly focused laws while the European Union defines personal information more broadly in its regulations. It is best to be familiar with your country's specific legislation if you use data that contains any personally identifiable information. Chapter 7 addresses these laws in greater detail.

Beyond national and international law, you will likely find organizational policies that cover sensitive data. These policies usually refer to national laws and therefore often deal more with the practical aspects of managing sensitive data. If you work for a corporation, however, be aware that most of your research data could be labeled sensitive for economic reasons. On a side note, the benefit of having a workplace policy is that there is sometimes a person responsible for making sure everyone follows that policy. You should view this person as a resource for helping manage your sensitive data.

There are several other types of data privacy policies worth noting. For instance, you should be aware of your local Institutional Review Board (IRB) policies if you conduct human subject research. Surprisingly, data sharing policies can also be useful in considering data privacy. Such policies often contain exceptions for sharing sensitive data and therefore define the data considered to be sensitive. Finally, any contracts you sign to use a third-party dataset often come with restrictions on storage and access to keep data private. Refer to Chapter 7 for more information on managing sensitive data.

3.3.3 Data retention policies

Data retention policies cover how long data must be actively maintained after the completion of a project. This is often in the range of several years to a decade, with the clock usually starting when the project/funding ends. The purpose of such retention policies is to ensure that data outlives the project and to make sure that data is available should questions arise. Indeed, there have been many cases (Oransky 2012; Oransky 2013b; Oransky 2013c; Oransky 2013d) where important data could not be found, either because it was lost or never existed at all, leading to investigation and retraction. Therefore, it is best to keep your data on hand post-project for as long as you can because you never know when you might need it again (Zivkovic 2013).

Funding agencies are the most common source for retention policies, though government funding agencies sometimes defer to umbrella government policies on data retention. For example, in the United States, the National Institutes of Health (NIH) has a three-year retention period (National Institutes of Health 2012) that echoes the retention period set by the US Office of Management and Budget (White House Office of Management and Budget 2013) for all federally funded research. Another source for retention policies is your local workplace. Be aware that if your institution has a required retention period, it may be buried within other data policies. Additionally, it is not uncommon for a local policy to state that data must be retained for "a sufficient amount of time" instead of giving an explicit number. When in doubt, I recommend a retention period of at least five years, and preferably ten years, after the completion of a project unless you have clinical, patent, or disputed data. These latter types of data usually require longer retention. Refer to Table 3.2 for examples of data retention times from a number of different institutions.

You should be aware that different types of data may require different retention lengths. For example, the Wellcome Trust in the United Kingdom states that the Trust "considers a minimum of ten years to be an appropriate period, but research based on clinical samples or relating to public health might require longer storage to allow for long-term follow-up to occur" (Wellcome Trust 2005). Besides clinical data, data supporting patents and data from studies involved in legal disputes or misconduct investigations must be preserved for a longer period of time. Patent data, in particular, has a long data retention period. You should preserve patent data and supporting notes for the life of the patent, usually around 20 years, in the event that you need this information during a patent dispute. Chapter 9 covers data retention in more detail.

Table 3.2 Selected data retention policies

Source	Country	Minimum retention	Starting point
National Science Foundation (NSF)	USA	3 years	"From submission of the final project and expenditure reports" (National Science Foundation 2013)
National Institutes of Health (NIH)	USA	3 years	"From the date the [financial report] is submitted" (National Institutes of Health 2012)
Engineering and Physical Sciences Research Council (EPSRC)	UK	10 years	"From the date that any researcher 'privileged access' period expires or ... from last date on which access to the data was requested by a third party" (Engineering and Physical Sciences Research Council 2013)
Biotechnology and Biological Sciences Research Council (BBSRC)	UK	10 years	"After the completion of the research project" (Biotechnology and Biological Sciences Research Council 2010)
Harvard University	USA	7 years	"After the end of a research project or activity" (Harvard University 2011)
University of Oxford	UK	3 years	"After publication or public release of the work" (University of Oxford 2012)

3.3.4 Data ownership policies

I'm going to start right out by saying that data ownership is often ambiguous. Research requires many resources, ranging from money to time to physical space, and it is often the case that these resources come from different sources. Many of these sources, be they people or institutions, have a stake in ownership but not everyone is a data owner. We therefore need policies that help us define ownership so we can properly manage data.

Lest you think data ownership is an abstract issue, disputes in ownership can have very real consequences to the progress of your research (Marcus 2013a; Marcus 2013c; Marcus 2013d; Oransky 2013e). One notable example of this comes from chronic fatigue syndrome researcher Judy Mikovits, who in 2011 was arrested for possession of stolen property: her own research notebooks from her former laboratory

(Cohen 2011; Schwartz 2012). The exact details of the case are somewhat sensational (including contamination, a retraction, arguments between her and the private laboratory where she worked over a clinical test for chronic fatigue, and a graduate student secretly removing notebooks and other materials from the laboratory), but the crux of the issue was that both she and the laboratory claimed ownership of the research materials. In the end, she returned the notebooks to her former employer, meaning that she no longer has access to her own work. While this is an extreme case, it shows that data ownership is a critical issue for managing data in the long term and determines what happens to research materials when a researcher leaves their place of work.

The first place to look for an ownership policy is with whoever is providing the money for research. Federal funders do not usually exert ownership on data, per se, but they may claim intellectual property rights, depending on the terms of the funding. This is less common in the United States, where the Bayh-Dole Act (Anon 1980) grants intellectual property rights from federally funded research to universities, small business, and non-profit organizations. However, it is best to examine the terms of funding to determine data ownership, especially if you are not publicly funded. After that, look for policies from your workplace. Corporations usually have the clearest expectations, in that the corporation owns any data you collect while working there. Ownership in academic settings is less clear, with many institutions lacking ownership policies and others defaulting to either university data ownership or ownership by the lead researcher. Finally, be aware that such policies may also take the form of contracts, either between research partners or researchers and their places of work.

Complicating this picture are collaborations, especially those that occur across institutions, as data ownership policies rarely cover such cases. This leads to confusion as to which member of the project owns the data – unless you sign a contract that specifically outlines ownership. I strongly recommend discussing ownership, storage, and access before you collect data for any collaborative project. This will prevent conflicts around ownership later in the project.

Finally, the person with the least likely ownership claim is frequently the one actually collecting the data. This is often a graduate student or technician. There have been several cases where a graduate student independently published data and was later forced by their adviser to retract the paper because he or she did not own the data (Marcus 2012; Marcus 2013b; Pittman 2012). More interesting is the case of Peter Taborsky, who filed a patent on work he did as a graduate student on using cat litter to treat sewage, a project that his adviser and corporate sponsor previously abandoned (NPR 1996). The university sued Taborsky for "stealing an idea" – no matter that he developed the technology independently – since Taborsky used university facilities for his work. Taborsky ended up serving time in jail for refusing to turn over the patent. The ultimate lesson here is that you should not automatically assume you are the data owner even when you collected the data.

3.3.5 Data and copyright

Like data ownership, data copyright is ambiguous, though in a different way. In contrast to data ownership policies, copyright is clearly set by national legislation and relevant court cases. For copyright, it is not the many possible sources of policy that create confusion, rather the interpretation of policy.

Confusion arises from the issue that facts are not copyrightable, yet creative compilations of facts, like databases, can fall under copyright. Adding to this confusion are the variations in copyright regulations from nation to nation. For example, the UK has a specific database right (Anon 1997) while in the United States, regular copyright only sometimes applies to databases. The best solution to this problem is to look up the copyright regulations for your country and to consult with your local copyright expert. More detail on copyright and data can be found in Chapter 10.

While copyright is important if you will use someone else's data for your research, you may also encounter licenses or contracts that can overrule any explicit copyright. Licenses define what permissions you have over the data and are either stated alongside the specific dataset or are given as a blanket license to all datasets from one source. Common licenses that allow data reuse include Creative Commons (CC) (Creative Commons 2014) and Open Data Commons (ODC) (Open Data Commons 2014). Unlike licenses, contracts are individual arrangements for the use of one or a group of datasets. Contracts are often used for data containing sensitive information or data with economic value. It's usually apparent when you are dealing with a contract for data reuse, but for data without a contract, it is worth looking to see if it falls under a specific license. Using openly licensed data is preferable to unlicensed data, as permissions are clear and you can avoid the often nebulous question of whether this dataset falls under copyright in your country. Again, refer to Chapter 10 for more information on data licensing and contracts.

3.3.6 Data management policies

One of the reasons you may be reading this book is because your funding agency has a policy on data management. Currently, the majority of these policies require you to create a data management plan as part of a grant. For example, the National Science Foundation in the United States requires a two-page maximum data management plan with every grant application. While such policies do not offer much advice in the way of how to actually manage your data, it's usually easy to find the expectations for the data management plan in the grant materials. Section 3.2 also offers suggestions on actually writing a data management plan and the rest of this book provides a range of strategies you can use for actual data management.

Beyond funders, many institutions have overarching data policies that cover data management, retention, privacy, and ownership. Unfortunately, like funder policies, these institutional policies are weak on the specifics of data management; most policies simply state that you should care for your data. Thankfully, such policies are usually helpful in other areas.

3.3.7 Data sharing policies

Data sharing policies have recently become more common and are being adopted by a variety of organizations. These include research funders, journals (Strasser 2012), and even whole scientific communities. The point of such policies is to increase reproducibility and, through data reuse, stimulate more research without more funding. Data sharing has also been shown to increase citations for the data's related article (Piwowar and Vision 2013). See Chapter 10 for more reasons for data sharing.

Funder and journal data sharing policies usually require you to share any data that forms the basis for a publication, with the exception of sensitive data. Funders often ask you to explain during the grant application how you will eventually share the data, or "make it available". Journals operate a little differently, as some journals request data be submitted along with a manuscript for peer review. Other journals want you to share data at the time of publication, even designating a particular data repository for sharing. You should consult with your journal and funding agency policies to learn the specifics of how to share your data; see Table 3.3 for examples.

Table 3.3 Select data sharing policies

Source	Conditions
Wellcome Trust	"The Wellcome Trust expects all of its funded researchers to maximise the availability of research data with as few restrictions as possible." (Wellcome Trust 2010)
National Institutes of Health (NIH)	"Investigators are expected to share with other researchers, at no more than incremental cost and within a reasonable time, the primary data, samples, physical collections and other supporting materials created or gathered in the course of work under NSF grants." (NSF 2013)
PLOS	"PLOS journals require authors to make all data underlying the findings described in their manuscript fully available without restriction, with rare exception." (Bloom 2013)
Nature	"Authors are required to make materials, data and associated protocols promptly available to readers without undue qualifications ... and [datasets] must be provided to editors and peer-reviewers at submission." (Nature Publishing Group 2006)

Community norms for data sharing are often more specific than journal and funder policies but less binding. An example of community norms are the Bermuda Principles (Marshall 2001; US Department of Energy Human Genome 2014), created by a group of geneticists, which require researchers to release all genome data from the Human Genome Project to the public within 24 hours of data collection. Community norms not only make you a good citizen of the community, but they are also great ways to learn how and where your peers share their data. For example, knowing that your peers put their data in a particular repository means putting your data in that repository makes it more likely to be findable. As my best advice for sharing data is to make use of existing sharing platforms, examining community norms can help you determine the best home for your data. See Chapter 10 for more guidelines on sharing data.

3.4 CASE STUDIES

The next two subsections contain examples of data management plans, the first a generalizable data management plan and the second my personal data management plan for this book. I hope you use them, in addition to the many other examples in this book, to identify ways in which you can apply the principles of data management to an individual project.

3.4.1 Example data management plan for a Midwest ornithology project

My main research data will consist of bird counts and observational notes, specifically for heron in the Great Lakes region of the United States. I will supplement this information with weather data, counts for other animals, and additional data as necessary. In particular, I will use counts from previous heron studies that are available via the Dryad data repository (www.datadryad.org) and through private sharing. While I will not maintain this external data, my bird counts and observation are unique data that will require careful management.

In total, I expect my data to be no more than 5GB in size for the entire three-year study. The data will primarily consist of Excel files (.xlsx), Word files (.docx) and .pdf files used for digitized observational data, and several R scripts for analysis.

Most of my observation documentation will be done in a field notebook. However, one of my collaborators, Dr. Joyce at the University of Rochester, will be using an electronic notebook to manage all of the data across the project. This will serve as the central location for the project data, though each collaborator will also keep a local copy, and as a way to organize everyone's data and notes. The basic organization scheme of the electronic notebook will be by project, by researcher, and then by date. I will use the file naming convention "YYYYMMDD_site_species.xlsx" for my observation data and "YYYYMMDD_site_species_notes.docx" for my notes. These conventions will cover both my local copy and the shared copy of my data.

I will store all of my data on my local computer and back it up automatically every day to a departmental server. After key points in data acquisition, clean up, and analysis, I will place a third copy of my data and copies of my notes in the e-lab notebook hosted by Dr. Joyce. I will retain ownership of my data, though Dr. Joyce will aid with storage, preservation, and access for other team members.

My data and code will be shared, in accordance with funding requirements. I will use Git (see Chapter 6) to version my analysis code and a GitHub repository to publicly share my R scripts. For the data, I will use the Dryad repository to share raw animal counts and supporting material, such as a data dictionary. One of my assistants will be responsible for preparing the data for sharing and I will be the main contact for questions. Data will be shared within one month of study publication.

At the completion of the project, I will keep copies of my data and code on Dryad and GitHub, respectively, and I will maintain my own copy of the data. I will be responsible for preserving all of the data, even after research assistants leave the project. Long-term storage will focus on two major areas, keeping up with new storage media and keeping up with file formats. To combat the latter, I will back up my data to the open file types, .txt and .csv, at the end of the project. I will also maintain a .pdf copy of my scanned field notebook.

3.4.2 My data management plan for this book

It will probably not surprise you to know that I created a data management plan at the start of writing this book. The purpose of this plan was mostly to allay my fears of a computer crash leading to total loss of my work. The secondary purpose was to organize my files so that I would be able to find everything at the end of the project. Since writing a book represents a huge effort and generates a lot of materials, managing content throughout the writing process was critical.

I started my plan by examining the major topics in this book and deciding which were most necessary for my information:

- Data documentation
- Data organization
- Storage and backups
- Long-term preservation

As I will not be doing data analysis or using another's data, I ignored these topics for my data management plan. Additionally, my data has no major security concerns, meaning that I took no extra precautions beyond normal storage and limiting access. Finally, my publisher controlled the collaboration systems used to prepare and edit the manuscript, meaning that I did not have to worry about information sharing as part of my data plan.

From this list of important topics, I created the following data management plan. This plan was easy for me to follow because it outlined many conventions that I made routine as part of my writing process.

File organization

My digital files will be organized in the following folders:

- "Outline": all files relating to the outline of the book
- "Draft": all copies of the book draft, broken down by pre- and post-edits and then by chapter
- "Final": final manuscript and versions of all supporting files
- "Documentation": content from my publisher and any miscellaneous documentation
- "DataManagementPlan": all documents for data management planning
- "TransferFiles": files transferred from other computers to my main writing computer

File naming
Chapter drafts will be named with the following convention: "ChXX_vXX.docx" (e.g. "Ch03_v12.docx" and "Ch10_v04.docx"). All other files will be descriptively and concisely named.

Versioning
I will periodically save chapter drafts and other major files as new version numbers, as outlined by my naming convention. I will work to keep versions up to date between my storage and backups.

Reference manager
I will use Mendeley (www.mendeley.com) to organize my references and aid with citations. I will periodically export my references in the .ris format to be stored in the "Drafts" folder under the "ReferenceLibrary" subfolder. The exported files will be named using the following convention: "YYYY-MM-DD_Library.ris" (e.g. "2014-01-28_Library.ris"). I will regularly back up these files in conjunction with my other documents and a copy of my current library will be cloud hosted on the Mendeley servers.

Storage and backups
The main copy of my files will live on my laptop. I will keep a backup copy on a local external hard drive and I will transfer files to this drive by hand at the end of each day. I will keep a second backup copy offsite using the encrypted cloud storage provider SpiderOak (www.spideroak.com); backups for this system will occur automatically.

Transferring files
When working on multiple computers, I will utilize SpiderOak's "Hive" folder to transfer files. I will copy items from the Hive "TransferFiles" folder to my main "TransferFiles" folder, which will be backed up according to my normal schedule. All files in both "TransferFiles" folders will be named with the date preceding a descriptive file name (e.g. "2014-01-02_ToDoCh04.docx").

Documentation
I will use README.txt files to document my files, as needed. I will keep one README.txt at the top level of the project, briefly describing the project and outlining several of the conventions laid out in my data management plan, including: file organization, file naming, and where the files are stored and backed up. I will also utilize README.txt files in each chapter draft folder to keep track of changes between chapter draft versions.

Long-term preservation
At the completion of the project, I will convert my files into more open or common

file formats, preferably .txt, .csv, and .tif. I will keep copies of both the original and the converted files.

Once the book is published, I will make an effort to keep track of where my files are stored and backed up. Whenever I get a new computer, I will copy the entire collection of book files from the old to the new computer. I will also keep two backup copies, one on an external hard drive or CD and one in secure cloud storage, being sure to check and update the storage media every few years. I plan to retain my files for at least ten years, preferably longer.

3.5 CHAPTER SUMMARY

Planning for data management takes time and effort but there are many rewards to putting data management systems into place before you do any research. To make a plan, you must consider the individual details of your data and research workflow and create solutions that fit your data in as many of the topics on the data roadmap (see Chapter 2) as necessary. Consult later chapters in this book to find data management ideas in each part of the data roadmap.

Once you decide on the plan for your data, you should write everything up into a data management plan. In general, your plan should address the following five general topics, but always defer to specific requirements when writing a data management plan for an outside source.

1. What data will you create?
2. How will you document and organize your data?
3. How will you store your data and, if necessary, keep it secure?
4. How will you manage your data after the completion of the project?
5. How will you make your data available for reuse, as necessary?

Even if your research funder requires the creation of a data management plan, I also recommend having a more extensive master plan for personal use. This is especially beneficial if you work in a group, as it allows everyone to use the same data management strategies.

Remember to consider relevant policies when creating your data management plan. These policies come from a variety of sources and cover a variety of topics, many of which are likely to apply to your data. Keeping in compliance with such policies will hold you in good standing with your institution, funding source, journal, etc. and can help you avoid legal entanglements.

Data management planning, in addition to being required for some types of funding, is useful for thinking through how you will handle your data during and after a project. Planning ahead makes it easier to document, organize, secure, and retain your data. Finally, a data management plan is useful for those who share data publicly, with collaborators, or even with their future selves, as it provides a roadmap for preparing data for reuse. So even if you are not required to create a data management plan, it is always a good idea to have one.

4

DOCUMENTATION

In 2013, a psychology article examining the correlation between shame and the desire for money was retracted for reproducibility issues. The corresponding author gave a summation of the problem: "my research assistant (who has left my lab last year) did not keep the original questionnaires. Hence, we cannot have correct data to rerun the data" (Oransky 2013g). Basically, without the original questionnaire, the authors did not have the proper documentation to verify their results. Sadly, this is just one of many instances where significant research problems arose due to lack of documentation.

When people ask me how they can manage their data better, one of my top answers is by improving their documentation. Anyone who has ever had the misfortune to go through their older data or someone else's data understands why. It is incredibly frustrating to try and use data that does not have adequate documentation (I say this from personal experience). In many cases, it's not even worth keeping data without proper documentation.

Documentation is one of the most important parts of managing data because data needs context in order to be understood and used. In some sense, data without documentation has no meaning. You wouldn't save a genetic sequence without stating the species to which it corresponds nor record the change in sample concentration over time without identifying the chemical that is reacting. All of these details are important. Therefore, you should think of documentation as a letter to your future self so that you understand what your data is and how you acquired it.

Besides helping you use your data, good documentation is also a very practical consideration. You are likely to forget the small details unless you write them down at the time. Too often has a scientist lamented "why didn't I write this down when I collected the data?" Don't let that be you! When in doubt, record these extra bits of information when you have access to them or are thinking of them. You never know when some trivial detail may prove important to your analysis.

Finally, documentation is fundamentally about helping yourself with your research but it also exists for the sake of others. Most often, these others include your co-workers and your boss. Documenting for your co-workers is a key reason to record extra information, as you may need to pass on your notes and data if you become sick or leave your current job. Regularly sharing information may also be part of your job, such as if you have a close collaborator or publicly share your data. Collaborators always appreciate a well-documented dataset, especially as it means they won't be pestering you with questions about the data. And of course, sometimes the people needing your data are auditors or those investigating research misconduct. The best way to prove your research was conducted properly is through good documentation. In all of these cases, it benefits everyone if you keep thorough and precise documentation.

The goal of this chapter is to provide you with ways to improve your documentation and make it easier for you to use your data because it is well documented. To do this, we'll cover best practices for research notes and notebooks, methods, other useful documentation formats, metadata, and standards. The idea is that you can mix and match documentation types to find something that works for you and your data. But no matter the format, always remember to keep sufficient documentation on your research data.

4.1 RESEARCH NOTES AND LAB NOTEBOOKS

Research notes have been the backbone of scientific research for hundreds of years. We can still examine Charles Darwin's writings to see his ideas on evolution take shape and Michael Faraday's meticulous notes on electromagnetism in his research notebook. In their hands, the simple nature of these unstructured notes allowed them to make great discoveries. Between the flexibility and rich history of the research notebook, this type of documentation has been, and will continue to be, an important part of scientific research.

While the main trait of research notes is that they are unstructured and can vary from researcher-to-researcher and even from day-to-day, there are some useful conventions to improve your note taking. Laboratory notebooks, in particular, have well-defined protocols for how to record content. This section covers all of these best practices, in addition to highlighting a new format for research notes – the electronic laboratory notebook.

4.1.1 Taking better notes

There are some general practices you should follow to keep good research notes, no matter the container or format of the notes. Good research notes have the foremost quality of being thorough yet succinct. You should record everything from observations, to the order in which you perform a complicated task, to the ID number and source of a reagent – you never know when spurious results will be linked back to a contaminated bottle of solvent (Stokes 2013). This level of detail must be balanced by notes that do not ramble. Don't let important details get lost in wordy passages. You want to make it as easy as possible for your future self or someone else to pick up the notes and figure out what you did.

To help with clarity, break up your notes and employ headings. One option for this is to follow the format of a research report or journal article by using the headings: introduction, experiment, results, discussion and analysis, and conclusion. Use subheadings as needed. Using tables, drawings, and figures will also add clarity and organization to your notes. Realize that some types of information are not presented well as a block of text, so experiment with different formats to make the most sense of your notes. Finally, label values and figures with the correct names and units.

When handwriting notes, write them legibly. If you are a researcher working in a foreign country, check that you are recording your notes in the appropriate language. Nothing is more surprising to a supervisor than looking over the notes of someone who just left the lab, and realizing that they are written in a language the supervisor cannot read (Kanare 1985). Remember that while your notes are primarily for you, they should be readable and understandable to others.

Finally, celebrate successes and be honest about mistakes. Any errors should be corrected by striking a line through them and noting the updated information, which preserves the original information in the event that it turns out to actually be correct. Your research notes are a vital part of your research and are the home for your observations and ideas. For this reason, you should be comfortable with the note-taking process and make it work best for you.

4.1.2 Laboratory notebooks

Taking good notes is not enough by itself to ensure that your documentation is useful. For this reason – and for the purpose of creating a trail of evidence for patenting – many other practices have been developed for laboratory notebooks. These additional practices, presented here as guidelines for keeping a laboratory notebook, can transform good notes into useful ones. I recommend adopting as many practices as you can to get the most out of your research notebook.

Keeping a laboratory notebook

The laboratory notebook serves as a record of the research process, including not only what you did but also your ideas, discussions, and analyses of results. The idea is that your notebook should be a thorough record of your science, so complete that someone "skilled in the art" (i.e. someone with similar training) could understand and reproduce your work simply from your notebook. In practice, this means that your lab notebook should be the hub around which your research occurs.

Most researchers use their notebooks to record the context of their experiments, writing down those little details that will be necessary for the analysis. Best practices, however, dictate a more complete notebook containing the following types of information:

- Raw data
- Documentation on experiments, including descriptions of set up, drawings and photos, reagent IDs, etc.
- Worked-up data, including graphs and figures, and notes on analysis

- References to the literature
- Cross-references to other pages in the notebook
- Research ideas
- Discussions about your research, such as email correspondence and conference presentations
- Table of contents and/or index

By recording this range of information in your notebook, the notebook becomes a true record of your thoughts and work during the research process. We can still go back to the notebooks of Darwin and see his thought process as he teased out the theory of evolution and this should be the case for any scientist's notebook.

Many corporations have clear policies and procedures for their laboratory notebooks and, where applicable, researchers should follow those policies. For researchers without such clear-cut policies, I lay out laboratory notebook best practices here.

New notebooks

The laboratory notebook itself should be a hardbound notebook containing high-quality, acid-free paper. These qualities prevent loss of pages and degradation of notes over time. If the notebook pages do not come numbered, this should be done by hand before using the notebook. Always use permanent ink, preferably a black ballpoint pen (which is less likely to bleed or fade), to write in your notebook (Kanare 1985).

When starting a new notebook, record your name and professional information, the date, and a summary of the notebook's purpose on the first page. This summary should include relevant co-workers, location of related information (like digital files and sample storage), the project funder, and a reference to any previous work done on this project. You can use the next one to two pages to record any abbreviations or organization and documentation conventions you will be using throughout the notebook. After these summary pages comes the notebook table of contents. Your table should have at least three columns for the date, a brief description of the record, and the page number of each notebook entry. A good rule of thumb is to allocate one row for each page in the notebook, which means setting aside a few pages in your notebook before your first entries. Regular notebook entries start after the last page allocated for the table of contents.

Adding notebook entries

To make a record in your notebook, first mark the date at the top of the page. Dates should be written in a consistent fashion, such as "Fri 1 Jan 2010", to make it easier to scan through the notebook. It is also good practice to record a brief heading for the day's tasks, the project name, and the name of anyone working with you at the top of the page.

When recording in the notebook, your writing should be clear and legible. Mistakes should not be erased or marked out. Instead, draw a single line through the error; the original writing should remain legible in case it turns out be correct after all. It is also advisable to add your initials, the date, and a brief description of the error for greater transparency. To aid with clarity, utilize headings such as "Experiment", "Data", or "Discussion" to organize information and use tables where appropriate.

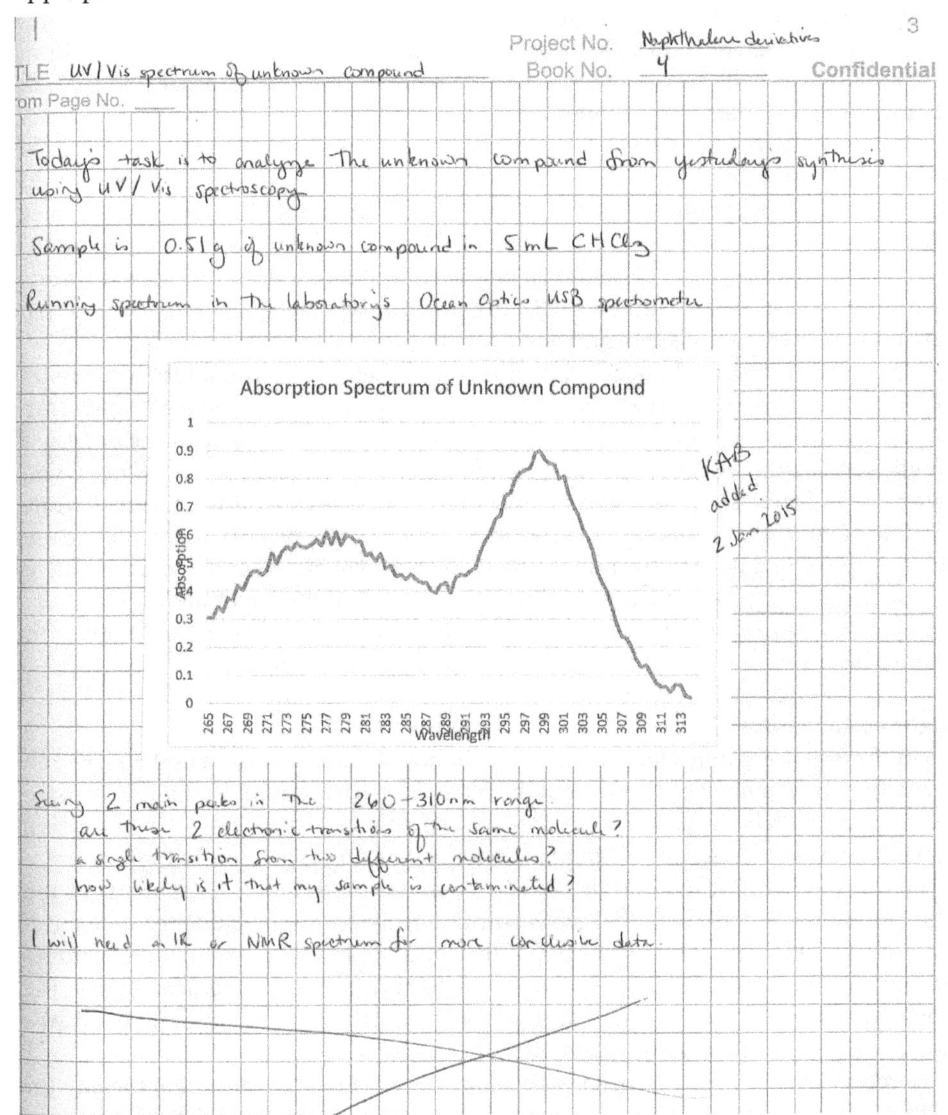

Figure 4.1 Sample laboratory notebook entry

In addition to notes, your notebook should include your research data. Unfortunately, adding data to your notebook can be complicated, especially because data is very often digital and/or large in size. There are three practical solutions to this problem.

The first option is to print out and paste data, or related graphs and figures, into your notebook when they are smaller than your notebook page. Use a good-quality acid-free white glue or archival mending tape to ensure these papers remain affixed to your notebook and do not yellow over time (Kanare 1985). Alternatively, print your data out and collect it in a three-ring binder, making reference to their location in your notebook. This option is also recommended for papers that cannot fit within the notebook as a single, unfolded sheet. Remember that this external information should be signed, dated, and witnessed if that is a requirement for your regular notebook pages. The final option for data, if printouts are not feasible, is to write data to a CD or DVD. One or two disks can be easily tucked into a paper sleeve and placed in the back of the notebook. Otherwise, keep your data stored in a central place and make a note of that location in your notebook. All of these practices developed under a paper-based system and therefore do not work well for moderate to large amounts of digital data. Refer to Section 4.1.3 on e-laboratory notebooks to learn how lab notebook software solves this problem.

Data printouts are not the only paper material that you should add to your notebook. In the course of work, researchers often write quick notes on a piece of scrap paper, such as while weighing ingredients at a separate workbench. If the notes are small, such as the weighed value, they can be copied in your notebook once you return to it. Some scraps, such as the elaboration of a research idea you had while outside of your work area, should be added to your notebook as is; follow the procedure for pasting data into your notebook. You should sign and date the scrap across the boundary of the scrap paper, making it obvious that the paper has intentionally been added and should not be removed. Figure 4.1 shows an example of how this should be done.

At the end of the day

If you finish taking notes for the day and still have space at the end of the page you are using, draw a line or "X" through the blank area to designate that it is intentionally empty. Notebooks are not supposed to be changed after the initial recording session and the practice of "X"-ing out space ensures transparency in that regard.

Get in the habit at the end of the day of updating the table of contents at the front of your notebook with that day's entries. Entries should include the date, brief description of the record, and page number(s). Keeping this information accurate and up to date makes it easier to quickly find information as you actively use the notebook as well as when you come back to the notebook in the future. This is one notebook practice that can save you time on a regular basis.

Some notebooks will also need to be signed by a witness at the end of the day (or within a few days) if the record is deemed important. The witness should read over the work for understanding prior to signing, so that they may attest "this page read and understood by me". This or a similar phrase should appear in the notebook with the witness's signature and the date.

Finally, notebooks should be stored in a secure place, such as in the laboratory or a locked desk drawer. Notebooks should not be taken home, as they might get lost.

Auditing notebooks

While a notebook is in active use, it is good practice for supervisors to periodically check their subordinate's work to ensure that the work is legible and understandable. The supervisor should make sure that all of the necessary notebook policies are being followed, such as for storage and witnessing, as well as checking that the researcher's notes are thorough and clearly laid out. This check should also include the table of contents to be sure that it is up to date. Supervisors should look over notebooks periodically to avoid the risk of discovering incomprehensible notes several years into the researcher's work.

Completed notebooks

Once you completely fill a notebook, make sure that the initial pages (summary, abbreviations/organization scheme, and index) are up to date and all of the necessary information has been added to the notebook. At this point, you should prepare your data for long-term storage – refer to Chapter 9 for guidelines – and make a copy of the notebook. You can make a physical photocopy but a digital scan of the notebook is often preferable. Scan to .pdf and save the notebook copy alongside the notebook's digital data. No matter the format, physical or digital, be sure to keep a copy of the notebook in a separate location from the original. This reduces the risk of loss in the event of disaster, fire, chemical spill, etc. at one location.

4.1.3 Electronic laboratory notebooks

If you are using a paper laboratory notebook, it is worth considering if an electronic notebook can better serve your needs. While the idea of a digital notebook has been around for a while, actual electronic laboratory notebook installations are finally becoming more common within scientific research. The pharmaceutical industry was one of the first major adopters of e-notebooks, with others in industrial research soon following suit. At this point, e-lab notebook technology has developed to the point that many laboratories, including academic and government laboratories, are going paperless.

The pros and cons of electronic notebooks

In considering the standard paper notebook, it is easy to see how best practices quickly break down when encountering digital datasets. It is no longer possible to add every data point or calculation to one's notebook or to organize all parts of your research using paper alone. This digital–analog divide makes it especially difficult to find information because you effectively must search twice for an experiment – once in the notebook and once on the computer. Not only does paper hinder searching, it is also fragile. Research notes are vulnerable to spills and to fire, not to mention the fact that they are difficult to back up, meaning that notes are often irrecoverable if the original notebook is lost. Finally, paper-based notes are more difficult to share in an era of increasing research collaboration. All of these issues together do not mean that paper notebooks are not a useful system, only that there are problems that an electronic notebook should address to make the transition from paper to electronic worth the effort.

Electronic notebooks do solve many of the issues presented by paper-based systems. Electronic notebooks allow for embedding files within the research notes, making it easier to find necessary information as you get the efficiency of searching digitally as well as finding everything with one search. Searching is made more robust through the ability to link between pages; instead of simply referring to a protocol outlined earlier in the notebook, the researcher can add a hyperlink to the relevant page. In general, electronic notebooks also track changes and who made them, even going so far as defining which people have read and/or write access, and to which content in the notebook. Other electronic notebook features make it easier to conduct research. For example, notebooks can sometimes be configured to read data directly from an instrument and even run tests to determine if the recorded values are within the proper parameters. Overall, there are many useful electronic notebook features but be aware that not every notebook will have every feature.

For all of the upsides to adopting an electronic laboratory notebook, there are also several drawbacks. The first is simply that the notebook is digital, making it susceptible to bugs, upgrades, security concerns, etc. Some of this risk can be reduced with the proper set up, though you may still be limited by the fact that you need computer, and possibly internet, access to use your notebook. Another concern is that there is not yet a clear best option for the software or file format. E-lab notebook companies come and go and there are little to no standards between one product and another. Because your notebook should not depend on someone else's business case, I cover exit strategies later in this section. The final major drawback for e-lab notebooks is the personal barrier. It takes time to transition from paper to electronic and not every notebook's user interface will work for every researcher. Then there is the reality that some people will always find paper preferable. Do not underestimate the effort of getting others on board as part of the adoption of an electronic-laboratory notebook. All of these things are concerns about adopting an electronic notebook, but many feel that the benefits of going digital outweigh the problems of staying with paper.

Basic electronic laboratory notebook functionality

If you decide to adopt an electronic laboratory notebook, you will first need to determine what exactly you need from your electronic laboratory notebook. At the very least, look for an electronic notebook with the following features:

- Robust note-taking capabilities
- Search capabilities
- Ability to embed data and image files and, preferably, view them within the notebook
- Ability to link between pages for referencing
- Secure log-in and tracking of who adds information to the notebook
- Audit trail that tracks the nature and time of changes in the notebook

- Ability to add witness signatures and lock notebook pages
- Ability to export to a common file type, such as a .pdf

The most important thing on this list is that the electronic laboratory notebooks provide good note-taking support. That said, the last point on exporting your notebook, what I call having an exit strategy, is important enough to merit further discussion.

Having an exit strategy

As electronic laboratory notebook software is still relatively new, it is very important that your electronic notebook should provide an exit strategy. While paper notebooks are accessible and readable well into the future, provided the paper is well cared for, the same is not true for an electronic laboratory notebook. Without access to the proper software and hardware, you do not have access to your notes. Additionally, there is currently no standard file type for exchanging information between notebooks, meaning that your older notes are likely to not be readable if you switch to a different notebook platform. For this reason, you should have a way to export notes from your electronic notebook in a usable format. Do not adopt an electronic notebook if you do not have a method for retrieving your data!

The time to ask about data export is while you are choosing notebook software. The best notebooks will export data in a common, open format. Look for .pdf, .xml, or .html, though .docx and .txt are acceptable but not ideal options. Notebooks that write to only obscure and proprietary file types are to be avoided at all costs. At the very worst, the notebook should provide a way to print all of the pages, as this is basically equivalent to a paper notebook. If printing is your only option, look for software that offers a batch job for this so you do not have to manually print every page.

Realize that you will likely lose functionality when you export. A collection of .pdf files will never have the same richness as an electronic record with linking and embedded datasets. However, the collection of .pdf files makes your notes readable in 20 years in a way that is not guaranteed by the software alone. Do not let your research notes be beholden to a business's success or failure – determine a way to export your notes before you even adopt an electronic laboratory notebook.

Other electronic notebook considerations

Beyond the basic requirements and having an exit strategy, you have a lot of options for an electronic notebook. Do you want open source software or something supported by a dedicated corporation? A local installation or software based in the cloud? A notebook with a lot of customizability or something tailored to your discipline? Do you need other features, like drawing tools and a mobile version? These are all choices for your electronic laboratory notebook.

Another important consideration in picking a notebook software package is how the notebook integrates into your existing workflows. There will be some adjustments, owing to the transition from paper to electronic, or from one software to another, but you should still feel comfortable using the software as part of your normal work. The best way to ensure that you like the user interface is to actually test out the notebook for a week or so. You will want to keep two notebooks during this period, as this allows you to walk away from the test notebook without any consequences to your record keeping. Do not underestimate the value of a good user interface; an electronic laboratory notebook could have all of the features you want but if they are not easy to use, you probably do not want to adopt that software.

Finally, if you do not find software you like or you are overwhelmed by the number of options available, wait a few years. Electronic lab notebooks are still adapting and settling into a stable marketplace. Eventually there will be some clearly preferred software packages and, hopefully, standards for file formats so you can still access your notes after switching software packages. Everyone must set their own balance point for risk between adopting software too soon and reaping the benefits of keeping electronic notes.

4.2 METHODS

Science loves protocols. Scientific research is built on the principles of consistency and reproducibility and it needs protocols to ensure those high standards are upheld. By following a tightly defined list of steps, researchers ensure that the results are the same each time or, when reproducibility is not possible, that all variables are accounted for. Science isn't science without protocols. But no matter how much these protocols shape our data, a protocol isn't data. Instead, protocols are an example of "methods".

4.2.1 Definition of methods

Methods mainly answer the question of "How?" Methods describe how to acquire a dataset, either under a particular condition or in a particular setting. They tell you how to prepare the sample, to culture the cell, or set up the apparatus. Methods also outline how you prepare data for analysis, which includes filtering and cleaning data, reconstruction, and the removal of artifacts. Methods can also be the code used for data analysis. Finally, methods tell you how to interpret the data, for example by stating how data points are grouped, information is coded, and what units the data is in. The overall theme is that methods piece together the history of the data but are not the data itself.

Scientists use methods information not only for reproducibility but also to build trust in data, sometimes called "provenance". By knowing how the data was generated and processed, we can understand the limitations and biases that are inherent in that data. Such information is critical to reaching a strong scientific conclusion. Therefore, methods information is necessary for interpreting that data and must be managed accordingly. Just as you would not publish a paper without a "Methods" section, don't preserve data for the future without methods information.

4.2.2 Evolving protocols

One concern with methods such as protocols is that they can evolve over the course of a study. For example, a sample preparation procedure may change as a researcher tweaks her experiment. The evolution of protocols is a natural part of doing research, but it can make the reconstruction of a dataset or an analysis difficult if you do not specify the exact version of the protocol used. For this reason, I recommend employing version numbers on any protocols that change over time.

Versioning protocols requires being explicit about all changes. Whenever a protocol changes, you should record it under a new number and note the differences from the previous version. You should rename both the protocol and any digital file in which it is stored, using a notation that is logical but clear. Here are a few examples:

- Calibration1, Calibration1b, Calibration1c, Calibration2, …
- "SamplePrep_v1.txt", "SamplePrep_v2.txt", …
- AnalysisProtocol_2015-10-23, AnalysisProtocol_2015-11-02, AnalysisProtocol_2015-11-15, …

In addition to clear naming, your research notes should also be explicit as to which protocol and version you used. You may end up with several versions of one protocol but, by using consistent numbering in the protocol names and the protocol used, it will be easy to keep track of exactly what you did. See Chapter 5 for more information on versioning files.

4.2.3 Managing methods information

The major management concern with methods information is that you remember to save and keep it with the data. There have been many sad cases where a researcher has examined a dataset only to find that they lack the necessary methods information, such as the original survey instrument, to usefully interpret it. You can either keep your methods with your notes or with your data, but most importantly, they should not be independently stored documents, which may be lost over time. Consider digitizing your protocols by scanning or retyping them and saving them to the data folder when wrapping up a project. Alternatively, rewrite the necessary protocols when you start a new research notebook. In either system, you should explicitly reference the protocol name or page number/date if the methods are recorded in your notebook. The ultimate goal is for it be obvious what method you used and also where to find it.

The secondary concern for methods is being sure to include any necessary units in your information. There is a huge difference between 1 gram and 1 kilogram, and 5 kcal and 5 kjoules. These are the small details that can make data useless if not properly recorded.

Finally, don't forget that methods often require their own documentation. For example, you should identify what the method is for, and under what conditions it should be used. Ideally, this would be recorded inside the method itself. A few notes describing the method can make a huge difference in your ability to use and understand it later.

4.3 OTHER USEFUL DOCUMENTATION FORMATS

Research notebooks and methods may be the most common formats for documentation, but some of my favorite documentation formats are the README.txt files, data dictionaries, and other structures covered in this section. That is because each of these formats fills a particular gap that is not well covered by notes or methods. Additionally, while most of these formats are simple, they pay large dividends in terms of use, search, and reuse. In short, these are the documentation formats to add to your repertoire in order to manage your data better.

4.3.1 README.txt files

One of the problems with the new abundance of digital data is that, often, research notes and research data live in two different locations. This means that data can be hard to use and understand because there is no nearby documentation to help with interpreting the data. Enter my favorite non-standard documentation structure: the README.txt file. These simple text files live alongside your digital research data to provide support for navigating folders and understanding the contents of your files. A good README.txt file makes your data easier to find and use.

README.txt files originated with computer science but are also useful for research data. In both cases, the README file is the first place to look in order to understand what the files are and how they should be used. The name itself implies that the file contains important information that should be read first. Additionally, the open file format, .txt, ensures that these files remain readable well into the future. These simple strictures make README.txt files an easy way to convey important file information.

README.txt files describe the general contents and organization of files and folders instead of the scientific information about them. You therefore use these files differently than your proper research notes. There are several types of README files that you can employ. The first is the README.txt at the top-level of a project folder. This file documents basic information about the project, such as:

- Project name
- Project summary
- Previous work on the project and location of that information
- Funding information
- Primary contact information
- Your name and title, if you are not the primary contact
- Other people working on the project

- Location of data and supporting information (lab notebooks, procedures, etc.) for the project
- Organization and naming conventions used for the data

This information allows someone to pick up your data and immediately know its purpose. Similar content should also be recorded in the front of your research notebook.

The other type of README.txt file can be used wherever you need it to add clarity within the project subfolders. These files generally answer the questions "What am I looking at?" and "Where do I find X?". For example, a mid-level README file may describe the organization of the files, as shown in the following example, or the laboratory notebook and pages that go with the data. A low-level README, however, might describe the technical arrangement of a set of data files. You do not need a README.txt file for every data folder, but you should use one wherever the data is not self-explanatory and you need clarity.

> This folder contains sample information from the Green Lake site, summer of 2009. This work was paid for by the NSF.
>
> Subfolders contain the following information:
>
> - "2009GL_Analysis": Chemical analysis data from all samples
> - "2009GL_Notes": Digital scans of the relevant field notes
> - "2009GL_Raw": Raw data
> - "2009GL_Weather": Temperature data from all days on site
>
> The file "conventions.txt" explains all naming conventions for the files in this folder.

README.txt files take very little time to create but are extremely useful for keeping your files organized and documented for the future. Create them as you generate and work with the data and also when you wrap up a project. It's worth spending a little time on this so that you can quickly understand your files when you return to them later.

4.3.2 Templates

If you keep research notes, you will want to consider using templates. Templates add structure to unstructured notes by creating a list of information to record for every experiment. Not only does this help with searching, but it also makes sure that you don't forget to record important information. A template is something small you can do to make your notes more thorough and consistent.

To make a template, take a few minutes at the start of a project to brainstorm the variables to record for each particular type of experiment you run. For example, a scientist might generate the following list for spectroscopy experiments.

- Date
- Experiment
- Sample
- Sample concentration
- Instrument power
- Wavelength
- PMT voltage
- Calibration file
- Measurement files

You will likely create a different list for your data.

Once you generate the complete list, post it in a prominent place, such as in your workspace, by your computer, or wherever you record your research notes. Every time you run that particular experiment, use the template to systematically record the necessary information. Having the list in front of you while you work will make sure you don't forget to record anything important. You can also type up this list and use it as a worksheet, but you must be sure that these pages do not get lost; they should either end up in your notebook or in a separate three ring binder that will always be stored with your notes. Finally, plan to reevaluate your list from time to time to ensure that it still reflects all of the necessary information.

4.3.3 Data dictionaries

Data dictionaries are my favorite documentation structure for spreadsheets and datasets containing many variables. This is because these documents allow someone to pick up a spreadsheet and quickly understand what it contains by giving context to the variables in the dataset or columns in the spreadsheet. What makes data dictionaries particularly effective is that this contextual information usually cannot fit within the spreadsheet itself. Data dictionaries give you space to define variables while keeping the actual dataset streamlined and computable. So if you have important spreadsheets that you plan to reuse or share, do consider making data dictionaries for them.

Generally, data dictionaries include the following types of information on each variable:

- Variable name
- Variable definition
- How the variable was measured
- Data units
- Data format
- Minimum and maximum values
- Coded values and their meanings
- Representation of null values
- Precision of measurement

- Known issues with the data (missing values, bias, etc.)
- Relationship to other variables
- Other important notes about the data

All of this information creates a framework through which to interpret a dataset.

To create your own data dictionary, simply go through each variable in your dataset and record what is known about it. Use the provided list as a starting point and add any extra information you think is useful. Your final data dictionary can go into a normal text file or README.txt file. Either way, be sure that the file stays with the dataset and is clearly labelled. Finally, it is a good idea to have someone else review your data dictionary and corresponding dataset to be sure the documentation is sufficient.

Using the basic outline, here is an example entry in a data dictionary for one variable in a dataset:

> Variable name: "PMT_cali"
>
> Variable meaning: Calibrated voltage measurements from the spectrometer's PMT detector. These values correlate to the sample concentration over time
>
> Units: millivolts
>
> Precision: PMT is precise to 0.002 mV
>
> Minimum: minimum value is set as 0 mV
>
> Other information: The PMT is calibrated daily, with raw values given in the "PMT_raw" column. Sample concentration calculated from the PMT values are in the "SmplConc" column of the spreadsheet

This example shows what the variable "PMT_cali" means and how to interpret its values. As part of a larger data dictionary containing information on every variable, such details would provide all of the information necessary to pick up and use a spreadsheet or multi-variable dataset. This is incredibly valuable for datasets that will be shared or are expected to have a long lifetime. The small effort to create a data dictionary is well worth it for the gains in clarity and context that this document can provide.

4.3.4 Codebooks

Codebooks are most useful when you are processing a lot of heterogeneous data into a consistent, computable dataset. For example, codebooks are extensively used on survey data to categorize responses. Such categorization requires subject knowledge, as you must first create the proper categories and then analyze your results to sort them into categories. The codebook itself explains both of these steps, the categories and the categorization process, and may also double as a data dictionary. This categorization information is critical because using different codebooks on the same raw data will result in two very different coded datasets. Therefore, you should create a codebook every time you code a new type of data so as to be explicit about the process of transforming raw data into coded data. Not only does this provide a reference point for you during the actual coding process, it is also useful to others for understanding how you coded the data. But remember, while codebooks are often paired with surveys, they are not surveys themselves. You should keep both the codebook and the original survey instrument with your data for analysis and preservation.

4.4 METADATA

While notebooks are standard in scientific research, one type of documentation that is increasingly being used in lieu of notes is metadata. Metadata, or "data about data", is an odd term but simply means information about your data. Technically, most types of research documentation fit under the broad definition of metadata but here I will use the term to denote a highly structured, digital form of documentation.

To really understand metadata, it helps to see an example. Here is some Dublin Core (Dublin Core Metadata Initiative 2014) metadata about a Western blot experiment:

Creator: Jane Larson

Contributor: Laura Wilson

Date: November 13, 2014

Title: Western blot from *Mustela erminea* samples

Description: Image of Western blot for samples from 2014-11-12. Protocol from 2014-02-05 with today's machine settings saved in the file "2014-11-13_BlotSet.csv"

Subject: *Mustela erminea*

Identifier: 2014-11-13_Blot01.tif

Format: TIFF image

Relation: Western blot image "2014-11-13_Blot02.tif"

The fields, or elements, are defined by the Dublin Core metadata schema while the user supplies the experimental information. This information can also be expressed as general research notes, though in a much less structured format:

Thursday November 13, 2014

Experiment: Western blot from *Mustela erminea* samples

Today I am running western blots on the samples L. Wilson and I prepped yesterday (see 2014-11-12 for sample information and 2014-9-23 for sample prep procedure).

All today's files will be saved to "C:/Documents/WBlots/2014-11-11/" folder on my lab computer.

I am using gel electrophoresis machine #3, with settings saved to file "2014-11-13_BlotSet.csv".

I am following procedure from 2014-02-05 for running these blots and using same control as from 2014-11-04.

<u>Lane – Sample number</u>

2 – MusErm_fm08

3 – MusErm_fm09

4 – MusErm_fm10

5 – MusErm_fm11

6 – Control

Image of blot saved as "2014-11-13_Blot01.tif".

The content in these two examples is the same, but the metadata version is much more structured than the research notes. This structure enables the metadata to be easily searched and mined, which is especially useful for researchers dealing with a large amount of data. So while you can often express documentation as either research notes or metadata, there are different advantages to each format.

4.4.1 When to use metadata versus notes

Using metadata during the research process requires different approaches than for research notes but comes with many rewards. The tight parameters of structured metadata make it the preferred format if you are searching through a large amount of digital information. What is a time-consuming visual scanning of notes for handwritten research notes becomes a simple text search using digital metadata. You can also perform calculations and comparisons on metadata that are not possible with other types of documentation. Another reason to use metadata is that metadata schemas provide a complete list of details to record for any experiment, observation, etc. This helps you manage and use data in the long term because you're more likely to have all of the necessary information in your documentation. Finally, metadata is an alternative to research notes and can capture the same types of information. So really, metadata is just another tool you can use to get the most out of your research data; use metadata when you need computability and research notes when you need flexibility.

Many researchers can benefit from using metadata instead of research notes for their documentation. In particular, researchers with large amounts of information to process should consider using metadata to facilitate searching and analysis. Collaborations can also benefit from metadata use. Heterogeneous data from different researchers is less of a problem when the metadata is consistent. Finally, researchers publicly sharing data are likely to encounter metadata in data repositories and databases. If you submit your data to a repository, be prepared to document your data according to the repository's metadata schema.

As digital data and documentation become more prevalent and larger collections of research data continue to be amassed, the use of structured metadata will only continue to grow. So even if you currently do not need to use structured metadata, it is worth knowing about it and how to adopt a metadata schema should the need arise.

4.4.2 The basics of metadata

The basic unit of metadata is a record for an object, which is usually a dataset in this case. Each record contains a controlled list of information about the dataset. The tightly defined record structure makes for easier searching over a large number of metadata records, and thus a larger number of datasets. The exact structure of a record depends on the metadata schema and the technology used to represent the record (called "syntax"), while the content of the record comes from the metadata creator.

Let's look at an example to see how this works. Here is a metadata record describing an observation of a loon.

```
<dwc:Occurrence>
  <dwc:basisOfRecord>HumanObservation</dwc:basisOfRecord>
  <dwc:individualCount>47</dwc:individualCount>
  <dwc:samplingProtocol>area count</dwc:samplingProtocol>
  <dwc:eventDate>2010-05-04</dwc:eventDate>
  <dwc:country>United States</dwc:country>
  <dwc:countryCode>US</dwc:countryCode>
  <dwc:stateProvince>Wisconsin</dwc:stateProvince>
  <dwc:locality>Mercer</dwc:locality>
  <dwc:decimalLatitude>46.17</dwc:decimalLatitude>
  <dwc:decimalLongitude>-90.07</dwc:decimalLongitude>
  <dwc:geodeticDatum>WGS84</dwc:geodeticDatum>
  <dwc:scientificName>Gavia immer</dwc:scientificName>
```

```
    <dwc:class>Aves</dwc:class>
    <dwc:genus>Gavia</dwc:genus>
    <dwc:specificEpithet>immer</dwc:specificEpithet>
</dwc:Occurrence>
```

This record uses the metadata schema Darwin Core (Darwin Core Task Group 2014) and the XML syntax. The schema provides the allowable elements, such as "dwc:basisOfRecord", and the syntax provides the format, generally "<element>value</element>". The metadata creator adds the proper information about the observation between the element tags. This researcher would create an individual record for each observation using the same schema and syntax.

Metadata schemas

A metadata schema defines the list of allowable content for that type of metadata. Many schemas exist for scientific research, each with its own disciplinary focus and allowable information. For example, Darwin Core defines its allowable elements as "dwc:basisOfRecord", "dwc:individualCount", "dwc:samplingProtocol", etc., which works well for descriptions of different taxa. Generally, you adopt a particular schema to fit your specific need.

Currently, most metadata schemas rely on element-value pairs where the schema defines an element and you list the corresponding value for the dataset in question. For example, a schema with a "location" element could have a value of "+43.9-110.7" for a particular record. Schemas can define a huge number of allowable elements, depending on the schema.

Beyond a list of elements for which you will define values, schemas usually contain:

1. Definitions of each element
2. Format for each value
3. Parent and child elements
4. Possible element qualifiers
5. Required and recommended elements
6. The number of times each element type may be repeated

For the fictitious "location" element, let us imagine how the schema could define all of these features:

1. Definition: "location" defines the place of observation rather than your workplace location
2. Format: use the format "+43.9-110.7" instead of "Jackson Lake, Wyoming, USA"
3. Parent: "location" is a child element of "observation"

4. Qualifiers: allowable qualifiers for the "location" element are "land", "air", and "water"
5. Required: you are not required to define a value for "location"
6. Repeat: you can list one value for location, maximum

All of this paints a very specific picture as to the meaning of "location" in this schema. And if this schema has tens or hundreds of elements, you can create a very detailed and structured description of your dataset. This in turn enables easier computation.

Metadata syntax

While the schema controls the content of a record, there are several ways to represent a record, referred to as the "syntax". The syntax governs the form of the information but does not change the content of the record. For example, you express records differently as a text file than in a database. Common syntaxes for structured metadata are plain text, XML (World Wide Web Consortium 2014; W3Schools 2014b), JSON (JSON.org 2014; W3Schools 2014a), and databases. Each syntax also comes with its own tools for metadata creation and searching. It is worth knowing a little about each of these syntaxes, as they all offer very useful tools for information management.

XML is stylistically similar to HTML and was designed as a data format for web applications. You will find many metadata schemas expressed in XML. Like HTML, XML stores data between matching tags and is hierarchical in nature, as shown here:

```
<observation>
    <location type="water">+43.9-110.7</location>
</observation>
```

JSON, which stands for the "JavaScript Object Notion", looks very different to XML. As its name implies, JSON works directly with the JavaScript programming language. JSON stores information as "element: value" pairs and can also incorporate some element relationship information, as shown:

```
{
    "observation": {
        "location": "+43.9-110.7"
    }
}
```

While fewer scientific metadata standards exist in JSON, it is a popular syntax for projects with a large programming component.

XML and JSON are both computer and human readable and have a discrete file for each record, but databases operate differently. Databases are made up of tables that contain multiple records, with one record per row and each column corresponding to a particular element. Databases are great tools for management of both metadata and data, particularly if you have large quantities of either. See Chapter 5 for more information on databases.

The choice of syntax is often limited by schema, as some schemas were created based in one or another technology. The schemas highlighted in Table 4.1 show the variability in metadata syntax, with some schemas supporting several syntaxes and others using only one. It is therefore recommended to choose a schema first, then learn the appropriate technology and tools.

4.4.3 Adopting a metadata schema

Once you realize that you want a structured metadata schema, you need to decide on the schema to use. A wide variety of scientific schemas already exists, so you should start by looking for existing schemas rather than creating your own schema. Creating a schema is time consuming and does not necessarily aid with sharing, so don't reinvent the wheel unless you must.

Table 4.1 Popular scientific metadata schemas

Metadata Schema	Discipline	Source	Description	Syntax
Dublin Core	General	(Dublin Core Metadata Initiative 2014)	General descriptive standard useful for a wide range of applications	Text, HTML, XML, RDF
Darwin Core	Biology	(Darwin Core Task Group 2014)	Adaptation of Dublin Core for biodiversity data, primarily taxa	XML, RDF, CSV
Ecological Metadata Language (EML)	Biology	(The Knowledge Network for Biocomplexity 2014)	Metadata schema for ecology	XML
Open Microscopy Environment-XML (OME-XML)	Biology	(The Open Microscopy Environment 2014)	Metadata for microscopy data	XML
Crystallographic Information Framework (CIF)	Physical Science	(International Union of Crystallography 2014)	File structure for crystallographic data	File format
Flexible Image Transport System (FITS)	Physical Science	(National Aeronautics and Space Administration 2013a)	Data format for astronomical images	File format

Metadata Schema	Discipline	Source	Description	Syntax
ISO 19115 and ISO 19139	Earth Science	(International Organization for Standardization 2003; International Organization for Standardization 2007)	Metadata standard (19139) and best practices (19115) for geospatial data	XML
Directory Interchange Format (DIF)	Earth Science	(National Aeronautics and Space Administration 2014)	Flexible metadata format for scientific datasets	XML
Climate and Forecast – Metadata (CF)	Earth Science	(Lawrence Livermore National Laboratory 2014)	Format for climate forecast model data as well as observational data	NetCDF

Table 4.1 lists several popular scientific schemas. Many other schemas exist, see the DCC list (Digital Curation Centre 2014c), so you must consider several factors in determining the best one for your project. First, will you be putting this data in a place that already uses a particular metadata schema? Many data repositories and databases utilize metadata schemas to aid with organization and searching and it will be necessary to utilize the repository's schema if your data will end up there. Another consideration is more informal collaboration and sharing: do your collaborators or peers use a particular metadata schema? Adopting the documentation standards of your field makes your data more understandable to others. Finally, what do you want out of your metadata schema? For subjects with multiple schema options, you should choose the one with a focus most aligned with yours. Librarians at your research institution can help you with this decision, as they are likely your best local authorities on metadata.

Once you identify a schema, look for tools that help with metadata creation, processing, and validation. I highly recommend creating digital metadata automatically whenever you can. Not only does this save you time by not having to type this information by hand, but automatically generated metadata is usually more consistent. Additionally, most metadata syntaxes come with tools for searching and analysis, like JQuery for XML and SQL for databases. Take advantage of these available tools to get the most from your metadata.

No matter the choice of schema, it is important to choose it early in the research process. This is because most schemas identify required information that, if it is not recorded when the data is acquired, is likely to be forgotten later or to have never been noted at all. Acquiring data with the schema in mind goes far in the prevention of incomplete metadata.

4.5 STANDARDS

A discussion on metadata naturally segues into one about standards, which are another way to add uniformity to documentation. Standards and their related brethren – ontologies, taxonomies, controlled vocabularies, categorizations, thesauri, and classifications – can be used in documentation for consistency in naming and in meaning. In general, standards operate by defining the allowable values on a particular topic, such as the list of natural amino acids, or the format for particular types of information, like describing location by latitude and longitude. So where metadata defines the variables to use, a standard provides the allowable values for a given variable. The inherent consistency of a standard aids in search and retrieval and also in sharing by defining how to interpret a particular piece of information.

The next sections provide an overview on several standards, though many others are available. At the very least, I recommend adopting some of the general standards such as for dates and units. It takes a little work to understand the qualifications of a standard but, once you do this, standards are a simple way to add consistency to your documentation.

4.5.1 General standards

While disciplinary standards are useful for certain projects, there are several general standards that are useful across all types of research. The first is ISO 8601 (International Organization for Standardization 1988; Munroe 2013), which covers date formats. This standard dictates that dates should be written using the following formats: YYYY-MM-DD or YYYYMMDD. So February 4, 2015 should be written as either "2015-02-04" or "20150204". In addition to this format, the standard provides several other formatting options:

- By day: YYYY-MM-DD, ex. "2015-02-04"
- By month: YYYY-MM, ex. "2015-02"
- By week: YYYY-WXX, ex. "2015-W06"
- By ordinal day: YYYY-DDD, ex. "2015-035"
- By date and time: YYYY-MM-DDThh:mmX, ex. "2015-02-04T14:35Z" (where X is the offset from Coordinated Universal Time, also called "UTC offset")

If you adopt no other standard, I recommend you adopt ISO 8601. The prevalence of dates in research notes means a standard date format makes searching simple. Additionally, using a standardized date at the start of your file names or in your spreadsheets makes it easy to sort chronologically and sift through information. It takes very little effort to write the date using ISO 8601 but you can reap many benefits from this small change.

Another ISO standard worth noting is ISO 6709, which governs the format of latitude and longitude. The general formats are as follows:

- ±DD.DD±DDD.DD/
- ±DDMM.MM±DDDMM.MM/
- ±DDMMSS.SS±DDDMMSS.SS/

Latitude always precedes longitude, with northern latitudes and eastern longitudes corresponding to positive numbers. So the latitude and longitude of Rome is given as "+41.9+012.5/". Note that the number of digits before the period is fixed – hence the padding 0 in the example – while you can use more or less digits following the period. Finally, this standard allows you to optionally specify altitude and the coordinated reference system used to interpret all the values. This standard's machine readability makes it useful for data analysis.

The final group of important standards is the seven SI base units:

- meter (m) for length
- kilogram (kg) for mass
- second (s) for time
- ampere (A) for current
- kelvin (K) for temperature
- mole (mol) for amount of substance
- candela (cd) for luminous intensity

These units each have very specific definitions, such as the "second is the duration of 9 192 631 770 periods of the radiation corresponding to the transition between the two hyperfine levels of the ground state of the cesium 133 atom" at 0 K (Taylor and Thompson 2008). This specificity means that these units add precision to scientific research. Many other units of measure exist, and several of them are even built upon the SI base units, but you cannot go wrong by using an SI base unit. They are so common that they are easily interpreted by most people versed in science, which is the whole point of using a standard.

4.5.2 Scientific standards

Scientific standards are useful in both data and documentation. For example, consistently using Linnaeus' taxonomy will make your data and notes easier to search and use. More often, however, these standards can help you find information in disciplinary databases. So even if you don't use them in your research, these standards, taxonomies, thesauri, etc. are useful to know about.

Biology

Biology is a field that loves taxonomies and standards, particularly where medicine is concerned. While medical standards are commonly used in medical records and reports, they also have value in documenting research and searching for information. It's mostly a matter of finding a standard that best aligns with your needs, as so many biological standards are available and there is often overlap between them.

Table 4.2 highlights a few of these standards, with more listed at the NCBO BioPortal (The National Center for Biomedical Ontology 2013) and the Open Biological and Biomedical Ontologies (OBO Foundry 2014) sites. Don't forget that your peers are also a good source to ask about standards, which aids with the sharing of information.

Table 4.2 Popular biological and medical standards

Standard name	Source	Description
Biological classification system	Linnaeus	Taxonomy of species and their relationships
Gene Ontology	(Gene Ontology Consortium 2013)	A controlled vocabulary for genes and gene annotation information
International Classification of Diseases	(World Health Organization 2013)	Diagnostic tool for epidemiology, health management and clinical purposes
MeSH (Medical Subject Headings)	(National Library of Medicine 2013)	A hierarchical system for medical terminology
National Drug File	(US Department of Veterans Affairs 2014)	Information on drug classes, clinical drugs and ingredients, and drug codes
NCBI Taxonomy	(National Center for Biotechnology Information 2013)	Database of information on all species with public gene sequences
SNOMED CT	(International Health Terminology Standards Development Organisation 2014)	A collection of clinical terms for human and animal medicine

Physical science

The major group of standards in chemistry relates to chemical naming. We usually teach students IUPAC naming (Favre *et al.* 2014; IUPAC 2014) for organic molecules, but there are several other chemical identification standards worth noting. The first two, InChI (InChI Trust 2013) and SMILES (Weininger 1988; Daylight Chemical Information Systems Inc. 2014), concern new ways of writing out chemical structures to facilitate data mining. Alternatively, Chemical Abstracts (Service 2014) and ChemSpider (Royal Society of Chemistry 2014) have their own unique chemical IDs to facilitate searching for specific chemicals in their respective databases. All four standards have very different forms than an IUPAC name, see Table 4.3 for an example, but all are useful in particular contexts.

Table 4.3 Different representations of the molecule acetone

Standard	Value
Common name	Acetone
IUPAC	Propanone
InChi	InChI=1S/C3H6O/c1-3(2)4/h1-2H3
SMILES	CC(=O)C
CAS	67-64-1
ChemSpider	175

You will also find several standards relating to physics and astronomy. The first is the Astronomy Thesaurus (Shobbrook and Shobbrook 2013), which is maintained by the Australian National University. This thesaurus covers a wide range of current astronomical terms and, instead of defining them, outlines their relationship to other astronomy terms. The NASA Thesaurus (National Aeronautics and Space Administration 2013b), developed and maintained by NASA, is similarly a thesaurus but also has occasional definitions. This thesaurus lists the subject terms NASA uses to index its reports and other materials, and covers engineering, physics, astronomy, and planetary science, among other subjects. Finally, those searching the physics and astronomy literature may wish to employ the Physics and Astronomy Classification Scheme (AIP 2014), developed by the American Institute of Physics, to aid in searching. While this classification scheme is no longer being updated, it has been used to classify fields and subfields of physics for the last 40 years. All three of these standards are useful for identifying the proper terminology and to aid with searching and sharing research information.

Earth science and ecology standards

The United States Geological Survey (USGS) maintains several standards relating to earth science and ecology, including the USGS Thesaurus (US Geological Survey 2014e), which covers a range of disciplines relating to earth science, and the USGS Biocomplexity Thesaurus (US Geological Survey 2014d), which focuses more on ecology. Additionally, the USGS keeps lists of other thesauri (US Geological Survey 2014c), dictionaries and glossaries (US Geological Survey 2014b), and controlled vocabularies (US Geological Survey 2014a) relating to earth science.

For those who work with geographic information systems (GIS), I recommend the GIS Dictionary (ESRI 2014) maintained by ESRI, a leading GIS software company. The dictionary defines many terms relating to GIS software and analysis. A second useful geographic standard is the Getty Thesaurus of Geographic Names (The J. Paul Getty Trust 2014), which contains a standardized list of place names as well as defining relationships between them. While this thesaurus was originally created for use in art and architecture, it may also be useful for scientists.

Math and computer science standards

Two classification schemes useful for indexing and searching in computer science and mathematics are the Mathematics Subject Classification (Society 2014), developed by the American Mathematical Society, and the ACM Computing Classification System (Association for Computing Machinery 2014), developed by the Association for Computing Machinery. In general, these schemes will be most useful in searching the literature, but may also have value depending on the project.

In computer science you are more likely to find ontologies developed for individual projects than other types of standards. Such ontologies are commonly used to add structure and logic to a project, making the project easier to implement. You can obviously develop your own ontology to fit your particular project but you may have the option of adopting or adapting a pre-existing ontology. This is true for both computer science ontologies and for ontologies in other disciplines – it is good to build upon previous work whenever possible, as it saves you time and effort.

4.6 CHAPTER SUMMARY

Research notes and lab notebooks are the standard documentation format for most scientists. While the format is fairly flexible, follow good note-taking and laboratory notebook practices to get the most out of your research notes. Additionally, consider moving to an electronic laboratory notebook if you have trouble finding information between your handwritten notes and digital data.

Methods information is a more specific type of documentation describing how the data was processed. Methods include things like protocols, code, and codebooks, all of which create the provenance of a dataset. The most important thing to know about methods is that researchers often forget to save and manage their methods, making the corresponding data difficult or impossible to interpret. So don't forget to hang onto your methods!

Beyond the common forms of research documentation, there are a few special documentation formats that are worth noting and adopting. These include README.txt files, templates, data dictionaries, and codebooks. All fill an important role and can help make your data easier to find, interpret, and use.

Metadata is a highly structured form of documentation that is increasingly being used in research. Metadata is most useful for those who must process and analyze large amounts of information and those who share data.

The final tools to improve your documentation are standards, thesauri, ontologies, and the like. General standards, such as ISO 8601 governing date formats and the standard SI units, are useful for most researchers. You should also examine standards within your discipline to aid with your documentation and to help you find relevant information in disciplinary resources.

5

ORGANIZATION

In 2014, a group of Swiss neurologists lost their paper "Spontaneous pre-stimulus fluctuations in the activity of right fronto-parietal areas influence inhibitory control performance" to retraction. The reason? "While applying the same analyses to another dataset, the authors discovered that a systematic human error in coding the name of the files" affected the article's conclusions (Marcus 2014). Basically, through poor organization and file naming, the authors accidentally analyzed the wrong files and came to the wrong scientific conclusions. This example of mistaken analysis is just one of the many ways that poor data organization can impact your research.

Organizing your files is a simple way to improve your data management. This is because it is easier to find and use organized information, to tell at a glance what work you've done, to wrap up a project, and to navigate your files well after the project is complete. Plus, with a little planning and some good habits, keeping your data organized is very achievable.

This chapter covers tips for organizing your data, as well as strategies like consistent file naming, using indexes and citation management software that will make it easier to find and use information. Many of these practices are simple enough to make a routine part of your research, which is the best way to enact change in your data management. These small things can make a big difference in dealing with your data.

5.1 FILE ORGANIZATION

File organization is simultaneously simple and difficult because it is conceptually easy but takes persistent work to have everything in its proper place. This is especially true in the laboratory, where data and notes pass between instruments, analysis systems, and even across the paper–digital divide. It is easy for things to get out of control and for files to spread. Still, all of the work required to keep files organized is worth it when you are looking for that one particular file that you really need.

The best piece of advice for staying organized is to have a system. The system should be logical, but more importantly, it should work well for you in your everyday research tasks. You must be comfortable with your system and make it a routine part of your work in order to stay organized. I describe several possible schemes in the following section and you should review them in order to adopt, adapt, or invent a system that works for you.

5.1.1 Organizing digital information

Digital information has the benefit of being somewhat easy to organize. You can create folders, copy and paste, and move everything around with the click of a mouse. But the ease with which you can manipulate bits also has a downside: it is very easy for those bits to propagate. A few copy-pastes across a couple of machines and you are left with information strewn everywhere.

The key with digital information is to be vigilant about staying organized. Pick a system for organizing your information and make a habit of putting data into the correct place. It also helps to have one main location for all of your data to keep your data from spreading across multiple machines. If you must work on data outside of the main storage, be sure to always place a copy in the central location when you finish your work. Alternatively, you can keep a list of places you store data outside of your central storage, but this means that you will have to maintain several storage and backup systems.

In terms of the actual organization of files in your storage system, there are many possible schemes to adopt for your data. A few ideas for organizing folders include:

- By project
- By researcher
- By date
- By research notebook number
- By sample number
- By experiment type/instrument
- By data type
- By any combination of the above

If you have a small project, you can place everything into a single folder, but most projects require folders and subfolders for organizing data.

The following example shows an organization system that arranges folders by researcher then by date. Date information is divided into two levels, by year and then by day, to further break up content. Also note that the lowest level subfolders sort chronologically because they follow the ISO 8601 dating convention (see Chapter 4). This system would work well for labs with many researchers who want to organize their data chronologically to match the arrangement of their written notes.

```
Mary
    2010
        2010-10-30
        2010-11-03
        2010-11-08
```

Continues

Continued

```
    ...
    2011
        2011-01-05
        2011-01-10
        2011-01-21
    ...
Kevin
    2012
        2012-05-13
        2012-05-28
        2012-06-03
    ...
    2013
        2013-01-07
        2013-02-03
        2013-02-05
    ...
```

There are many possible organization systems and your best system will depend on your research workflow and how you generate and search for data. Take a few minutes to brainstorm and design a system that works for you. Consider the homogeneity and heterogeneity in your data and how you might naturally group data. Also consider how you search for data; for example, regularly searching for data from a particular experiment versus a particular date suggests that it may be better to organize data by experiment than by date. Your system does not have to be complex and it's likely that everything won't fit into one perfect system. Still, any conscious effort to organize most of your data will help you keep track of files in the long run.

Finally, know that you can adopt a consistent organization system or even improve upon your current organization system mid-project. You can do this either by reorganizing all of your files into the new system or by leaving your old files as is and arranging only your new content under the new system. Because it is relatively easy to move digital files, it is often worth the effort to rearrange older files if you end up with a system that works better for you. However, this is not a good time investment in all cases. Researchers with a large number of project files, for example, should simply switch to a better organizational system and leave their old files in the previous system (being sure to document the change in their notes). In the end, you must decide if the effort of reorganizing older files is worth the time you can save later when looking for this content.

5.1.2 Organizing physical content

While physical samples and documents are very different as compared to digital files, the same organizational principles apply. Namely, that you should have an organizational system and stick to it. For physical content, it is best to come up with this organizational scheme at the outset as it can be difficult to reorganize content in the middle of collecting material. Refer to Section 5.1.1 for possible organizational schemes or just use the scheme that works best for your content.

In addition to having a system, well-organized physical collections also benefit from good labeling. Whether this is labeling folders full of printouts, dates in laboratory notebooks, or ID numbers in a drawer full of samples, you should have some way of announcing what things live in a particular location. For large shared collections, you can take this further by not only labeling content but also documenting the general organizational scheme to help others more easily use the collection. The general rule of thumb for organizing physical objects is to make it as easy as possible to find what you need at any point in time.

5.1.3 Organizing related physical and digital information

Most scientists live in a world with both digital and analog information, usually in the form of digital data and written notes. This is a particularly challenging place to exist. First, it's twice the work to maintain two separate systems, being sure to have adequate organization and backup for each. Second, it is difficult to find related information when some of it is digital and some of it is analog. Compounding this problem is that, often, scientists organize their written and digital collections in different ways; for example, arranging written notes by date and digital data by experiment type. While this type of system is often the easiest to use when acquiring data, it quickly becomes difficult to navigate the analog–digital divide when analyzing that data.

One obvious solution to the analog–digital problem is to switch to an entirely digital system. Electronic laboratory notebooks are becoming a viable option for research and such systems can manage both notes and data in one central location. This keeps everything together and makes for easier organization. See Chapter 4 for more information on electronic laboratory notebooks.

For those unable to switch to an entirely electronic system, you can use the same organization scheme for both your physical and digital collections. Organizing everything by date is a natural choice, especially if you keep a laboratory or field notebook, but you are free to pick the best organization system for your data. Using the same organization scheme makes for easier searching because if you know where your notes are in one system, you know where your data is in the other and vice versa.

In lieu of converting everything to one standard, you can also employ indexes. An index allows you to organize your digital and analog information in different ways and use a list to translate one system into the other. For example, if you organize your notes by date and data by experiment, you can build an index for your notes sorted by experiment and/or an index for your data sorted by date. This makes it easy to go back and forth between related analog and digital information. The next section goes into more detail about how to make and use an index.

5.1.4 Indexes

Indexes are a useful way to impose order on data without having to actually rearrange any files. They operate by creating a list of content ordered by topic and stating where the relevant information can be found. If you have ever used the index at the back of a book, you understand the value of using an index to facilitate searching.

> **Index vs table of contents**
> An index and its sibling, the table of contents, are different but related structures. The table of contents usually appears at the front of a book and lists items in the order they appear in the book. The index is usually located at the back of the book and lists contents in a different order, such as by topic in alphabetical order. Look at the table of contents and index in this book to see the difference between the two. While it's good practice to use a table of contents for your research notebook, you should also consider an index, particularly if you organize data and notes using different systems or regularly search for a particular type of content.

In the laboratory, you can use an index for your data, your notes, or anything else you want to find easily. Indexes are particularly useful for managing written notes and digital data, as it's much easier to use an index than to scan through two separate systems to find related pieces of information. Additionally, it is helpful to use an index for frequently searched for content, such as protocols or important digital files that may not be stored in the same location. Think of the index as a quick guide for finding what you need.

The simplest type of index outlines a particular type of content and where it lives. For example, an index could list every calibration file saved on your computer and its location. A second type of index identifies related information, such as files and notes, stored in different locations using different organization systems. The major difference with this type of index is that it orders a list of contents from one system by the organization scheme of the other system, effectively translating one organizational system into the other. For example, an index could list your digital files arranged to match the ordering scheme in your research notes, as shown in the example that follows. Both types of indexes are useful, depending on your needs.

Research Notebook Order	Index	Order of Data
1: Experiment A notes	1 = 3′	1′: Experiment D data
2: Experiment B notes	2 = 2′	2′: Experiment B data
3: Experiment C notes	3 = 5′	3′: Experiment A data
4: Experiment D notes	4 = 1′	4′: Experiment E data
5: Experiment E notes	5 = 4′	5′: Experiment C data

In order to create an index, the first thing to do is decide what information to index and how to order it. For example, if you arrange your digital data by sample number, you may wish to develop an index for your research notes organized by sample number. Other orders to consider are:

- By date
- By subject
- By experiment type
- By observation site
- By sample type
- By notebook page number
- By protocol type
- By any scheme logical for finding your content

Depending on the content and reason for which you are building an index, one organizational scheme is likely to be preferable to the others.

Once you decide on an arrangement for your information, consider what other information you may want in your index. Things like the date, the experiment, or a sample description add context to what is being indexed and the location of the corresponding information, making it easier to scan through your index. In the following example index of laboratory protocols, a description and date add useful information to the protocol type and its location.

Protocol Type	Description	Date	Notebook page
Sample prep	Mouse care	2010-03-17	5
Sample prep	Tissue prep	2010-04-29	34
Sample prep	Tissue staining	2010-07-14	108
Calibration	For laser	2010-03-21	10
Calibration	For microscope	2010-06-03	62
…			

With the outline of your index in hand, you need to populate the index with information. The easiest way to fill an index is to do it as you conduct your research. Get into the habit of updating the index at the end of each day and, with very little time investment, you will always have an up-to-date index. Alternatively, take time to scan through your notes or data and add the relevant information to the index. This is not a difficult task but it does take time and attention to detail.

Indexes and tables of contents can be handwritten, but consider maintaining a digital index. The benefit of using a digital index is that it is easier to sort and search for information. In particular, a spreadsheet-based index can be re-sorted on the fly, allowing it to be more comprehensive and useful under a variety of conditions. There are many options for creating an index but remember that an index is a tool that must work for you, meaning function trumps form.

5.1.5 Organizing information for collaborations

Staying organized is critically important when researchers collaborate and share information. Such collaboration exists under a variety of conditions: with an immediate co-worker, with peers at another institution, at a particular point in time, or building a data archive for the future. No matter the circumstances, you should think about organization when you collaborate on a project as good organization makes it easier to find and use others' content.

The best way to organize information for collaboration is to have everyone use the same system. This means using the same conventions for organization, file naming, and version control on a central storage platform. Being consistent about file organization and naming makes it easier to find project data, even if you are not the data creator. If you know the schemes, you know how to find what you need. Be sure to agree on conventions at the outset of a collaboration so everything stays organized as you acquire data.

An alternative to everyone using the same system is to have several organizational systems that are well documented. It doesn't matter how many people use one particular organization scheme, so long as it is understandable to other people in the collaboration. This means documenting conventions in a highly visible location so that collaborators can easily learn and leverage organizational systems. While this system is not as seamless as the previous one, it can be easier for individuals to generate data while still allowing for active collaboration.

Finally, if you use collaboration-enabling software – such as an electronic laboratory notebook – it's still worth using consistent naming and organization. Not only does this improve upon basic searching in the software, it also allows your files to be used independently should you stop using the software. Just because you have a useful tool doesn't mean that basic principles stop being important.

5.1.6 Organizing literature

The last topic of this section does not concern data directly but is still important to the research process: organizing your literature. This encompasses organizing the .pdf files saved on your computer and keeping track of the articles you read. You need slightly different approaches to manage both types of information.

Organizing .pdf files – indeed, any type of digital file – relies on many of the techniques outlined in this chapter. Both good organization and good file naming will make a big difference in keeping track of digital copies of articles. While the overall organizational structure is up to you, there are some things to look for in a file naming system for articles. Specifically, it helps to put citation information like author name, title information, journal, and/or year in the file name. For example, one system that works well is to use the first author's last name and publication year followed by the title – truncated as necessary – as the file name. This system gives file names such as:

- "Schmitt_2006_SynthesisOfFeSiNanowires"

- "Bard_2011_TetraoninaeBonasaNestingHabitsInAlaskanForests"
- "Lipp_1998_ProteaseXYZKineticsInChimpanzees"

This system makes it easy to find articles by author and browse for article by title, but it is just one of many possible naming systems. Ultimately, you should use the information you will remember best about an article or something that will spark your memory in the file name. Once you have a system in place, you must be consistent with how you name files to reap the benefits of a naming convention. See Section 5.2 for more information on naming conventions.

Beyond basic file organization, I highly recommend citation management software. There are many options currently available, including: EndNote (www.endnote.com), Mendeley (www.mendeley.com), Papers (www.papersapp.com), RefWorks (www.refworks.com), and Zotero (www.zotero.org). All of these software applications store citation information and provide you with the ability to add notes and manage your .pdf files. The real power of a citation manager, however, is its ability to create bibliographies on the fly in a wide range of formats. You don't need to worry about the details of MLA, Chicago style, or an obscure journal-specific format because the software does it all for you. You will want to use a citation manager if you read and write a lot of scientific articles.

There is one more thing to note about citation managers and that is that they allow you to save and export citation libraries. This is important for a couple of reasons. First, this is useful for collaborative writing so that all parties use the same set of literature as they write (note that you may also need to use the same citation manager for this to work best). Second, being able to save your library to an independent file allows you to properly back it up. It is easy to forget to include this information in your normal backups because many software packages store library files apart from data. As it's a setback to lose your reading library, be sure to periodically export your library and back it up with the rest of your data.

5.2 NAMING CONVENTIONS

Using consistent naming is one of my favorite tips for good data management. It takes the benefits of good file organization to the next level by adding order to the actual files within the folder. While these naming conventions nominally apply to files and folders, you can use these principles whenever you need to name groups of objects, such as physical samples. For how simple it is to adopt good naming, there are many rewards for using such conventions.

5.2.1 File naming

You should use consistent names for the same reason that you use good file organization: so you can easily find and use data later. Additionally, good naming helps you avoid duplicate information and makes it easier to sort through your data when wrapping up a project. Conversely, inconsistent naming can lead to disorganization but also incorrect analysis if you are not properly monitoring which files and samples you use, as seen in the example at the beginning of the chapter.

File naming conventions work best for groups of files, folders, or samples that are somehow related. These groups can be small (less than ten), but consistent naming is particularly useful for large groups of files. It's also likely that your data will not all fit neatly into one naming convention, which is fine. You can easily come up with several naming conventions for different groups of files and follow them once established.

Your biggest consideration for a naming scheme is what information should go in the name. Good names convey context about what the file contains by stating information like:

- Experiment type
- Experiment number
- Researcher name or initials
- Sample type
- Sample number
- Analysis type
- Date
- Site name
- Version number (see Section 5.2.2)

Basically, you want just enough information in the name to immediately and uniquely identify what is in the file. This will vary across and within projects, so do not be afraid to adopt more than one naming convention if that works better for your data. For example, your raw data might be named by site and date while the names of your analyzed data files list the analysis type and experiment number. Once you identify the most important content to describe your files, you should create a consistent system for naming your files.

You have a lot of leeway with naming systems, so long as you follow a few general principles:

1. Names should be descriptive
2. Names should be consistent
3. Names should be short, preferably less than 25 characters
4. Use underscores or dashes instead of spaces
5. Avoid special characters, such as: " / \ : * ? ' < > [ ] & $
6. Follow the date conventions: YYYY-MM-DD or YYYYMMDD

The first principle is that names should be descriptive. This means that you get useful information about a file solely from its name. For example, "MarsRock13_spectrum01.xlsx" tells you a lot more about the data than "mydata.xlsx". Second, names should be consistent. This includes not only the overall naming structure of related files but also how you code meanings into the name. For example, you might include the names of data collection sites and you should always write the site name (or abbreviation) the same way. Third, names should be short. Even if they are consistent, long names make it harder to find and sort through content. Fourth and fifth, you should avoid certain characters in file names, such as spaces and special characters. This is because some computer operating systems cannot handle these characters in file names; note that underscores and dashes are acceptable and good alternatives to spaces (refer to the inset box on camel case for further recommendations). Finally, I recommend using the date convention YYYY-MM-DD or YYYYMMDD (i.e. "2016-05-17" or "20160517" for May 17, 2016) if you want to add dates into your names. Not only is this a standard convention, ISO 8601 (covered in Chapter 4), it is especially useful at the beginning of a file name because files will sort chronologically.

Between basic principles and desired content, you should come up with a system for your files. Here are a few sample naming systems:

1. YYYY-MM-DD_site (date + site name)
2. YYYYMMDD_ExpmtNum (date + experiment type + experiment number)
3. Species-expmt-num (species name + experiment type + experiment number)
4. Expmt_Sample (experiment type + sample name)
5. YYYYMMDD_source_sample (date + sample source + sample name)
6. ChXXvXX (chapter + version)

Those systems translate into the following file names:

1. "2014-02-15_Plymouth" and "2014-03-17_Salem"
2. "19991021_WesternBlot07" and "20010516_WesternBlot02"
3. "Vlagopus-obsrv-112" and "Vlagopus-count-067"
4. "UVVis_KMnO4" and "IRspec_CH4"
5. "20040812_AppleLake_35cm" and "20050812_AppleLake_50cm"
6. "Ch04v12" and "Ch05v03"

All of these examples convey a certain amount of information just in their name. While this will not be all of the important information about a file, it is enough to easily sort through your data if you pick a good system. Of course, interpreting a file name hinges on the fact that you know the convention and relevant codes used in naming, so be sure to write this information down (see Section 5.3 on documenting conventions).

Once you have a system in place, make it a habit to name your files and samples according to your conventions. Naming conventions only work if you use them consistently, but thankfully they are simple enough to make a routine part of your research.

Camel case and pothole case

The field of computer science offers several useful conventions that we can co-opt for better data management. Two that are applicable to file naming

Continues

Continued

are "camel case" and "pothole case". These conventions eliminate the need for spaces in names – which most early (and even some modern) computer systems could not handle – while still allowing complex names to be readable.

Camel case creates order in names through a mix of uppercase and lowercase letters. It dictates that for a name containing several words strung together without spaces, you capitalize the first letter of each word and leave the rest lowercase. The very first letter in the name can either be upper- or lowercase. Examples: CamelCase, maxHeight, TempInKelvin, AlbuquerqueMeasurements, sample03Analysis.

Pothole case differentiates words in a name by separating each one with an underscore. The letters are often all lowercase, but are not required to be so. Examples: pothole_case, max_height, temp_in_Kelvin, Albuquerque_measurements, sample_03_analysis.

Both camel case and pothole case are useful in creating detailed yet readable file names, but I'm not so much of a purist that I'm above mixing the two in one naming system. I encourage you to play around with both cases in your naming conventions to come up with names that are descriptive yet readable.

5.2.2 File versioning

In addition to systematically naming your files, another helpful naming strategy is to include version numbers in your file names. This means that as you work on a complex analysis, write a paper, etc., you should frequently save to a new file with a new number. This simple system works well for linear changes that do not require complex tracking. For more robust version control systems, refer to Chapter 6.

Utilizing file versioning is a useful data management strategy for a number of reasons. First, it allows you to track your progress and easily revert to earlier versions of the file. So if you make a mistake in your analysis, it is easy to jump back a few steps; versioning means that you don't have to start over from scratch. Additionally, using version numbers is useful if you are working collaboratively on a file or are doing work in multiple locations. You can send different versions back and forth and always know which is the latest. I highly recommend versioning files when collaborating or working on complex processes.

To version your files, simply save the file at key points to a new version number. You decide when these key points occur and how often to spin off a new version. Distinguish different versions using a version number at the end of the file name; you can also offset the number with a "v", underscore, or dash. Alternatively, label versions with a date. It is also recommended to label the final version as "FINAL" instead of a number or date. This designates that the work is complete and the "FINAL" version is the one to use.

Here are a few examples of versioned file names:

- "PlasmaPaper_v01", "PlasmaPaper_v02", ..., "PlasmaPaper_v15", "PlasmaPaper_FINAL"
- "sample31Spectra1", "sample31Spectra2", ..., "sample31SpectraFINAL"
- "RefLib-01", "RefLib-02", ..., "RefLib-FINAL"

The downside to this very simple versioning is that it does not capture the differences between different versions. Therefore, it's a good idea to take notes on what each file version contains (or alternatively, use a more powerful version control system). You can do this in your research notebook, but it's usually better to keep these notes alongside the files in a README.txt (see Chapter 4). The next section covers many good ways to document both conventions and versions.

5.3 DOCUMENTING YOUR CONVENTIONS

Most of the work of implementing conventions is coming up with them and making their use routine. There is, however, another requirement for using a convention that should not go overlooked: documenting your conventions. Any conventions you come up with – be they conventions for organization, naming, or versioning – you should document. This allows others to gain an understanding of your files just from your documentation, which is necessary should you share data, fall ill and need assistance with your work, or plan to hand over your data when you leave a laboratory. Additionally, it's a good idea to document your conventions in case you forget them and want to use your files in the future.

5.3.1 What to document

To document your conventions, you should provide a basic framework for how the conventions work and note any information necessary to decoding your system, such as abbreviations used in files names. Documentation should be clear and concise, with as much detail as necessary to make a peer understand what you did. The following is an example of documenting organization and naming conventions.

> I use the following conventions for my datasets.
>
> ORGANIZATION
> Datasets are organized into two main folders: "RawData" and "AnalyzedData".
> "RawData" contains the master copies of the data, which is further divided into folders by collection site and then by date.
> Any changes to and analysis of the raw data are done on copies of the data in the "AnalyzedData" folder. Data in this folder is grouped by analysis type. README.txt files in each subfolder record which datasets are analyzed, noting both the site and date for the data.
>
> *Continues*

Continued

NAMING

Raw data files use the following naming convention: "SiteAbbr-YYYYMMDD-Num" (site abbreviation + date + sample number).

Examples: "BLA-20000710-2", "MAM-20010508-1", "EME-20010627-5"

I use these site abbreviations:

"AZU" = Azure Spring
"BLA" = Black Pool
"BOP" = Black Opal Pool
"EME" = Emerald Spring
"MAM" = Mammoth Hot Springs
"SUR" = Surprise Pool

Analyzed files use the following naming convention: "Analysis-Description-vX" (analysis type + brief description of data + version number of file). Analyzed datasets will be further described using README.txt files in relevant subfolders.

Examples: "AtomicEmission-FeContent-v2", "pH-SURsamples-v3", "Temp-200105data-FINAL"

In the course of analysis, files should periodically be saved to a new version with an updated version number; final versions should be labeled "FINAL".

It's always a good idea to have someone look over your documentation early in the project, particularly if you are collaborating and sharing data. This will help ensure that your documentation is logical and complete. Particularly in the case of collaboration, clear documentation will eliminate the need for collaborators to ask you about your organization and naming schemes anytime someone wants to use the data.

5.3.2 Where to document

Once you have a description of your conventions, you should put it in several prominent places such as your research notebook, on a print-out next to your work computer, in a file with your data, or anywhere you are likely to either see the rules or wherever it is logical to look for them. I recommend putting at least one copy with your written notes, like the front of your lab notebook, and one copy with any digital files, preferably in a README.txt file (see Chapter 4).

README.txt files are ideal for documenting conventions, especially those concerning digital data. You should place a copy of your overall conventions in a README.txt file at the top-level of your project folders. Additionally, if you have conventions that apply to a subset of your data files, you should document those in

README files in the applicable folders. Refer to Chapter 4 for more information on using README.txt files.

For those using laboratory-wide conventions, you should document and store them in a communal location in addition to keeping your own personal copies. For example, you could store a digital copy on a laboratory shared drive, taking care to place this file in a communal folder. A communal computer is another good storage location. Besides storing your conventions in a communal location, be sure to periodically review them as a group to make sure everyone is following the conventions and to update any practices, as necessary.

5.4 DATABASES

Databases are a great way to organize research data and are especially useful for searching and querying data. Many of the things you do on a day-to-day basis revolve around databases, such as browsing your favorite website (organizing website content) or making a purchase at the corner store (financial and product inventory systems). Databases, particularly relational databases, are so ubiquitous in the non-research world that it is worth considering how they can help in the research arena.

The power of a database comes from the fact that it is a highly ordered arrangement of a large amount of data. This rigid structure enables querying across the data at speeds not possible with almost any other system. Furthermore, queries can be very complex, ranging from basic calculations to searching for items that meet a unique set of criteria. For example, if you want to know if you observed more of species X or Y on Tuesdays when it rained, a good database will find you the answer more quickly than scanning through your notes.

Databases are ideal for when you are analyzing a large amount of data and for when many people need to query a collection of information. Metadata (see Chapter 4) is also well suited to databases because of a metadata schema's rigid structure. The downside of a database is that you must put your data into the database in a tightly controlled manner; this takes time and effort, though some automation is possible. A database is not going to be the solution for every research group, but is worth considering if you and/or your co-workers do a lot of searching and sorting through data or metadata.

Big data technology

With the explosion of big data, many in the business world are discussing relational database alternatives, the primary example being Hadoop. Such systems operate on an even larger scale than relational databases – working best for terabytes of data or more – because they are built on a fundamentally different model. The major difference is that Hadoop does not have the rigidity of a relational database and instead gets its computational speed

Continues

Continued

from using a large number of processors in parallel. There is obviously much more to Hadoop and big data technology, but I recommend skipping the hype around these systems unless you work with an unusually large amount of data. You'll save time and effort by sticking with a traditional database.

To better judge if you need a database, it might help to know how a database works. The next sections provide a basic overview of the structure within a relational database and how to query one. This should provide you with a general framework for understanding and using a database in your research.

5.4.1 How databases work

The majority of databases are relational databases. Unlike a spreadsheet, a relational database does not put information in one large table but is instead made up of many small, related tables. Using many tables leads to faster searching because the database does not have to sort through every piece of information during every search. Additionally, a relational database is not limited to the single table format with one and only one value in each cell. This makes a relational database more space efficient and allows the user to associate multiple values with any one variable for each entry in the database.

The actual arrangement of a particular database will depend on your data, as that determines the number and form of the tables. For example, if you perform blood tests on mice with multiple tests per animal, you could have one table containing test subject information, another containing the schedule for testing, and a third containing test results (see Table 5.1). A different project would have an entirely different table structure. Once you set the structure for your database, you can populate it with data. Columns correspond to different variables and rows correspond to entries.

Table 5.1 Example database consisting of three related tables

Subject Table

id	name	gender	age	type
12	Izzy	female	13	A
13	Prudence	female	30	B
14	Thor	male	7	A
15	Harriet	female	22	A
16	Max	male	16	B

Schedule Table

testNum	id	date	time	tester
123	13	2007-03-26	09:00	Beth
124	15	2007-03-26	13:40	John
125	13	2007-03-28	09:00	Beth
126	16	2007-03-29	16:25	John
127	13	2007-03-30	09:30	Brad

Results Table

testNum	bloodCt	FeLevel
123	279	53
124	288	39
125	291	53
126	342	48
127	306	53

What makes a relational database work is using unique IDs to associate entries across tables. The example in Table 5.1 shows how this works. There are two unique IDs in this example database: "id", which defines a unique ID for each test subject, and "testNum", which provides a unique number to each testing event. Two entries in the Subject Table cannot have the same ID, likewise for entries in the Results Table having the same test number. The association between the two IDs in this database occurs in the Schedule Table, which connects a unique subject ID to a unique test number in each entry of that table. Note that this arrangement also allows every subject to have multiple tests. Leveraging IDs enables us to break up data into tables and take advantage of relational database efficiency, so long as we are careful to make sure IDs are unique.

If you understand the concept of related tables connected with unique IDs, you have a solid foundation for understanding how relational databases work. This will allow you to create and use a database in your research. You must still design your database and populate it, and for that I recommend consulting one of the many database books currently in print, or a local database expert, as there are many individual database products available. Still, identifying that you need a database and understanding the basic database framework are the important first steps.

5.4.2 Querying a database

The second part of understanding databases is in knowing how to use them. This applies to anyone using a database, even if you didn't build it. Using a database means performing searches – called queries – on a database to look for patterns or extract particular values. Querying is very powerful and highly customizable, which means it is a deeper topic than can be covered in a small section of this book. However, understanding the general framework of a query will allow you to create useful searches and provide a foundation upon which to learn more complex querying.

Most databases use the Structured Query Language, or SQL (pronounced "sequel"), as the framework for searching the database. Even when databases do not appear to use SQL, it is often because the software places an interface between the user and the actual SQL commands. Either way, understanding the structure of a SQL query will enable you to write better queries, no matter the particular format.

A basic SQL query has the following structure:

```
SELECT [variable name]
FROM [table name]
WHERE [variable name] [operator] [value];
```

It often helps to read the query backward to understand what it means. For this generic query, the "WHERE" clause identifies the conditions you are querying for, "FROM" identifies the table(s) which stores the information, and "SELECT" defines what to return to the user from the set of entries that meet the specified conditions.

Let's examine a real query that uses this basic structure. This example queries the tables outlined in Table 5.1.

```
SELECT age
FROM Subject
WHERE gender = "female";
```

Reading this query backward, we see it looks for all females in the Subject Table. The database then pulls out the entire entry (including id, name, gender, age, and type) for each female in the table. The "SELECT age" command filters out most of this information, so that only the age values for this set of entries are reported to the user. So basically, this query returns the ages of all female test subjects, or [13, 30, 22].

Let's look at a more complex example. Here is a query that returns the name and IDs of all subjects tested by John.

```
SELECT S.name, S.id
FROM Subject S, Schedule C
WHERE S.id = C.id AND C.tester="John;
```

This query is more complex because it is done across two tables; we want to query for "John" from the Schedule Table but report back name and ID from the Subject Table. There are two things to note about how this query is different from the previous one. First, we must specify to which table each variable belongs by putting the table identifier in front of the variable. Second, we must connect related information between the two tables, as we only want to connect the entries in each table that correspond to one another. This happens by stating that the subject IDs in the Subject and Schedule Table must be the same. This query returns the following information: [(Harriet, 15), (Max, 16)].

The foundations of querying are logic and understanding the structure of your database, and you can use these to build on the basic query structure to do many interesting things with your database. There is obviously more to querying. From here I recommend picking up a reference book on SQL or referring to the W3Schools SQL Tutorial (W3Schools 2014c), which is a good reference guide for many details of the SQL language. Databases are extremely powerful for search and analysis and with a good foundation in the structure of querying you can unlock great potential through the analysis of research data.

5.5 CHAPTER SUMMARY

Keeping files organized is an easy way to improve your data management. Good file organization means having a system for organization and always putting everything in its proper place. Indexes can also help with organization by pointing to the location of particular content. Finally, using citation management software will help you organize your literature.

In addition to basic organization, consistent naming will help you find and organize your files and physical samples. This file naming can be extended to create a simple version control system to track changes to a file over time. Be sure to document any file naming and organization conventions in an obvious place.

A final tool for organization is the database, which can store data and metadata in highly ordered tables. Such a structure is useful if you have a lot of data and want to perform complex searching and analysis upon it.

6

IMPROVING DATA ANALYSIS

Errors in data are a part of doing research. Sometimes they pop up because you wrote down the wrong thing and other times they arise due to forces outside of your control. Take, for example, the case of dates in Excel:

> As it turns out, Excel "supports" two different date systems: one beginning in 1900 and one beginning in 1904. Excel stores all dates as floating point numbers representing the number of days since a given start date, and Excel for Windows and Mac have different default start dates (January 1, 1900 vs January 1, 1904). Furthermore, the 1900 date system purposely erroneously assumes that 1900 was a leap year to ensure compatibility with a bug in – wait for it – Lotus 1-2-3. (Woo 2014)

If you use Excel and port files into different analysis programs or simply open them in Excel using different operating systems, you will have to check for date errors. Now this is a particularly egregious example of data error but it, like all other types of errors, can slow down the analysis process. As the goal of data management is to always make it easier to work with your data, data management offers several strategies – such as quality control – that can help streamline your analysis.

Data management plays a smaller role during the data analysis stage of the data lifecycle, but there are still several data management strategies that can improve your ability to properly evaluate your data. This chapter covers managing raw versus analyzed data, strategies for quality control, spreadsheet best practices, coding best practices, and using version control systems for software. All of these practices aim to help you improve the analysis portion of the research process.

6.1 RAW VERSUS ANALYZED DATA

Before getting into the details of managing data during analysis, let us define the two types of data relevant to this portion of the research process: raw data and analyzed data. Raw data is the primary information you collect in order to perform your analysis. This type of data can be anything from the direct output of an instrument, to counts done in the field, to a dataset pulled from the internet that you intend to reuse. No matter the form or polish of the dataset, raw data is the original material.

Raw data is defined by its lack of processing.

Analyzed data exists in many stages, from cleaned up to semi-processed to presentable. Each of these stages may have value to you as a researcher, depending on what you want to do with your data. In particular, cleaned up data is an important stage for many researchers as this is the point where they can start the real analysis. Examples of cleaned up data include filtered data coming off the Large Hadron Collider (processed to make the data a more manageable size), reconstructed image files, and noise-reduced or averaged data. A second important analysis stage is for presentable data. This is data that is completely analyzed and goes into a table or is converted directly into a figure for disseminating your research results. Presentable data may not be directly published, but is the final form of your data, post-analysis.

Raw and analyzed data, being two different types of information, naturally require different management strategies. Additionally, the process by which you transform raw to analyzed data is also important to record and manage. Let us now look at the best practices for managing all of this content in more detail.

6.1.1 Managing raw and analyzed data

It's good practice to always maintain a copy of your raw data so that you can redo your entire analysis or test a different method. You never know when you may need to start over. Keep raw data in a separate location from analyzed data so that you do not accidentally save over the files and make a copy of the data before starting any new analysis. You can also make these files read-only as a precautionary measure. Finally, be sure to denote, either through the file name and/or in your documentation, that these are your raw data files.

While raw is the preferred form for this baseline copy of your data, it may not be the most feasible form to maintain. Depending on your data and your resources, you may wish to always retain the cleaned up – but still unanalyzed – copy of your data instead. The previous case of filtered data from the Large Hadron Collider is a great example of this, as it is not feasible to keep every terrabyte of data that comes off that instrument. In maintaining a cleaned up form of your data, you must be sure that your cleaning procedure is correct and no useful data is lost or damaged in the process. Maintaining cleaned up data is good practice, but only if it's not practical to keep a copy of the data in its raw form.

Besides keeping a master copy of your raw data, you may also wish to maintain copies of your data at key points in the analysis. The number and form of the copies depends on the project, but it's a good idea to keep multiple copies if you perform multiple analyses. The one copy of analyzed data you want to be absolutely sure to maintain is the presentable copy of your data, meaning the finalized version of your data used to generate figures, tables, and other portions of published results. Keeping a copy of this data on hand lets you address questions about your data and easily return to your data if you want to use it again in the future. As with raw data, be sure to note in the name and/or documentation that this is the final version of your data.

Finally, you should prioritize the management of any data, raw or analyzed, that is not easily reproduced. For example, observation data tied to a unique time and place requires more care than a dataset generated from a quick simulation (in the second case, the priority is often the management of the code). The more unique the data or time and effort required to obtain the data, the more you should take care to prevent data loss.

6.1.2 Documenting the analysis process

There is a third important type of information on the spectrum of converting raw to analyzed data and that is the process of analysis itself. Just as you would not collect data without noting the acquisition process, so too should you note the analysis process that takes your data from its raw to its final form. Such information includes everything from analysis procedures to the version of software (custom made or proprietary) used. Keeping track of your analysis makes it easier to repeat, easier to correct errors, and easier to describe to others when publishing your results.

There are several ways to keep track of your process. One is simply by taking good notes while you perform your analysis. These notes can be kept in your research notebook or in a file stored with the analyzed data. If you use custom code for your analysis, you should save your code alongside your analyzed data and, if the code has evolved over time, note the version used. Another option for working with code is to keep log files on the programs and commands used on your data. Analysis is rarely a seamless process, so try to capture as much information on the process as you can and automate this capture whenever possible. The ultimate goal of documentation is to allow someone else (or your future self) to repeat your analysis process on this or related data.

6.2 PREPARING DATA FOR ANALYSIS

Often, half of the challenge of analysis is preparing the data for analysis. This means cleaning the data, performing error checking and quality control, and making the data more consistent. This section focuses on many of the small practices that can streamline subsequent analysis. Like many things in data management, a lot of small practices can add up to a happier relationship with your data.

6.2.1 Data quality control

Quality control makes good data great and helps smooth out the rough edges during analysis. All data should undergo some sort of quality control, including checking for errors and ensuring consistency across the dataset. Not only will this make analysis easier, but it will also make it more accurate.

Data consistency

One of the best things you can do to streamline your data analysis is ensure that your data is consistent, meaning your data should have the same units and format across each variable. Consistency prevents errors by making sure your values are all correct. Do not forget the lesson of the NASA Mars Climate Orbiter, a spacecraft that was lost due to different parts of the instrument calculating related values in different units (Sawyer 1999). Beyond absolute correctness, consistent formatting makes for easier analysis because related values are all represented in the same way. Examples include using the same encoding for digital textual data and using the proper number of significant figures in numerical data.

Textual data is particularly difficult to keep consistent between variant spellings, synonyms, and formatting. If you work with any spreadsheets containing textual data (and even if you don't), I highly recommend using a clean up tool such as the program OpenRefine (www.openrefine.org). OpenRefine, a free program that used to be Google Refine, can identify related information and variants in spreadsheet data, allowing you to make the data more consistent, modify formatting, or discard particular data points. For example, you can use the tool to identify all of the variants of a plant name and change them all to the same Latin name or remove parentheses around any instances of a plant's common name. If you are dealing with a large amount of tabular data or data that contains multiple format errors or misspellings, plan to use a data clean up tool like OpenRefine.

Error checking

Beyond form and formatting, quality control means checking your data for errors. Such errors may be a typo, an instrumental error, or a value in the wrong unit, but you want to be sure to check for them before working your analysis. At its most basic, error checking means scanning through your data to look for improbable and missing values, though there are other methods for checking. Your preferred method will depend on the form of your data and the tools you have available, but here I offer two powerful ways to perform error checking.

One of the easier ways to check for errors is to make a simple plot of your data (or subsets of your data) along some logical coordinate. Outliers and artifacts in your data are often easier to identify visually than when buried within a large dataset. For example, Figure 6.1 shows a dataset with an outlier that may not be obvious until graphed. Another graphing option is to distribute your data into groups (or "bins") and create a histogram of the number of items in each group. The group divisions can be as coarse or as fine as you choose and will show you the rough distribution of your dataset, a useful check to determine if your data is as expected. Figure 6.2 shows an example of this technique. Not only does this plot tell you that the data has a Gaussian distribution, but it's also easy to spot the errors for several points in the middle. Both of these methods require little effort in processing and can help you easily identify anomalies in your data.

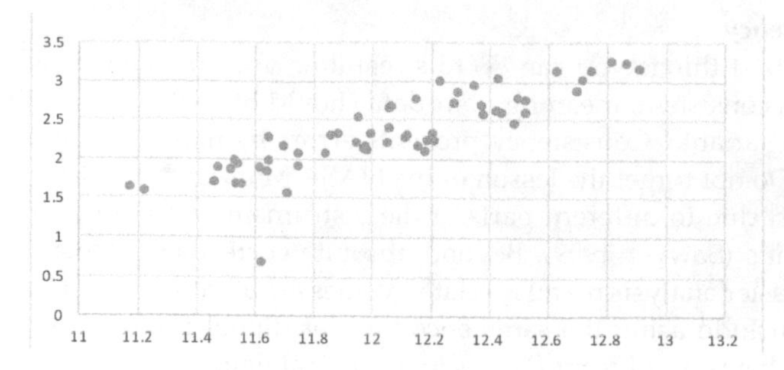

Figure 6.1 Dataset with outlier

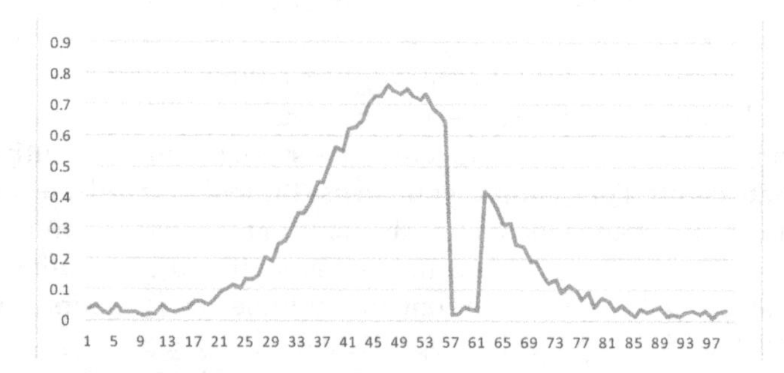

Figure 6.2 Dataset containing an error

Zero versus null

Another big problem with data arises when you want to represent the absence of value. This can mean many things including true zero, that no measurement was taken, and that the variable does not apply to this entry in the dataset. The problem occurs when we try to represent these distinct cases in a dataset. Often these cases are randomly assigned a label, with datasets sometimes ascribing one label to multiple meanings or multiple labels to one meaning. Commonly used labels include:

- NULL
- N/A
- 0
- -
- No data
- None
- [blank]

Clearly, there needs to be a more consistent way to represent absence of value in order to properly analyze data. In particular, there should be a distinction between values that are intentionally zero and values that are not.

The first rule is to only use "0" to represent true zero, meaning when you actually took a measurement and that measurement was zero. This will clear up confusion as to whether "0" represents an actual measurement or not.

For all other conditions, the label must denote that there is no intentional measurement. The best way to do this is to leave the entry blank, which most programs will interpret as a null value (White *et al.* 2013). The blank entry does not distinguish between missing and intentionally null data and you must take care not to use a space to denote a blank, but it is still the best option to denote lack of value across a number of programs. Individual programs will also accept different values for null ("NA" in R, "NULL" in SQL, and "None" in Python) but this is less optimal, as you cannot transfer datasets between programs without modification. So start with "0" and blanks as your default and add from there as needed, being sure to thoroughly document any variations in labeling.

Avoid compound variables

While not strictly a process for quality control, it's also good practice to avoid compound variables in raw datasets. Compound variables are variables calculated from other variables (called component variables), an example of which is body mass index (BMI) calculated from height and weight (NYU Health Sciences Library 2014). Storing the component variables, height and weight, not only allows you to calculate BMI but also allows you to calculate other values, such as BMI alternatives. You gain more flexibility in data analysis if you choose component variables instead of compound variables during your data collection.

The one exception to the recommendation against compound variables is for personally identifiable information. Compound variables can often be used to mask personal information without excluding useful data from your dataset. For example, while it may be more useful to store subjects' full birthdays, it is often better to store their age. Age gives you roughly equivalent information to birth date and is less identifiable. For researchers with sensitive information, you must consider the trade-offs in analysis and security when deciding on what variables to acquire in your datasets.

6.2.2 Spreadsheet best practices

Spreadsheets are a common way to analyze scientific data and serve as an important model for good practices in consistency and computability (Strasser 2013; UW-Madison Research Data Services 2014). In the framework of data management, spreadsheet best practices are likely to differ from the current way that you use spreadsheets in that they emphasize readability by the computer over human readability. This is mainly because computer-readable spreadsheets are easier to analyze, reuse, and port into a different data analysis tool. You shouldn't have to reformat, move, or delete cells in order to perform an analysis. Therefore, to make the most of your spreadsheet data you should view your spreadsheets more as data tables than as notebook pages containing both notes and data. Notes instead should go into a data dictionary or README.txt (see Chapter 4) to ensure that a human can still understand the data. Formatting your spreadsheets for computer readability is not difficult but does require some intention in planning your spreadsheets.

At its most basic, a good spreadsheet has no extraneous formatting. This means no merged cells, no highlighted or colored text, no embedded figures, etc. Merged cells and embedded figures prevent the computer from being able to easily analyze data and formatting adds information in a form that the computer cannot usefully process. Additionally, a good spreadsheet only contains one table; unrelated tables should be moved to separate tabs or even a different file. Figure 6.3 shows an example of a bad spreadsheet with formatting and multiple tables, with Figure 6.4 showing corrections to this spreadsheet. You should view your spreadsheet as a set of values organized into rows and columns, anything beyond which is likely unnecessary. A spreadsheet with minimal formatting can easily be exported for deeper analysis in R, on the command line, etc. So remember, all of the extra formatting may make the spreadsheet look good to you but it does not help make the spreadsheet do what is intended – data analysis.

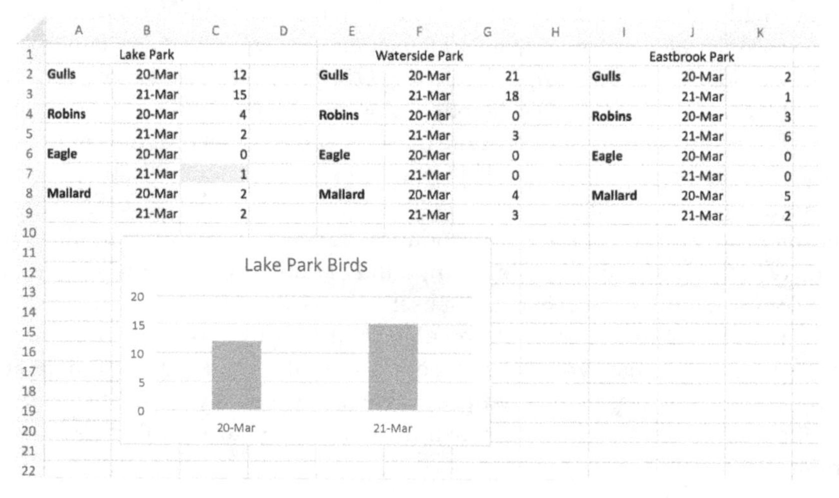

	A	B	C	D	E	F	G	H	I	J	K
1		Lake Park				Waterside Park				Eastbrook Park	
2	**Gulls**	20-Mar	12		**Gulls**	20-Mar	21		**Gulls**	20-Mar	2
3		21-Mar	15			21-Mar	18			21-Mar	1
4	**Robins**	20-Mar	4		**Robins**	20-Mar	0		**Robins**	20-Mar	3
5		21-Mar	2			21-Mar	3			21-Mar	6
6	**Eagle**	20-Mar	0		**Eagle**	20-Mar	0		**Eagle**	20-Mar	0
7		21-Mar	1			21-Mar	0			21-Mar	0
8	**Mallard**	20-Mar	2		**Mallard**	20-Mar	4		**Mallard**	20-Mar	5
9		21-Mar	2			21-Mar	3			21-Mar	2

Figure 6.3 Poor spreadsheet formatting with multiple tables

	A	B	C	D
1	Site	Date	Species	Count
2	Lake Park	20-Mar	Eagle	0
3	Lake Park	20-Mar	Gull	12
4	Lake Park	20-Mar	Mallard	2
5	Lake Park	20-Mar	Robin	4
6	Lake Park	21-Mar	Eagle	1
7	Lake Park	21-Mar	Gull	15
8	Lake Park	21-Mar	Mallard	2
9	Lake Park	21-Mar	Robin	2

Figure 6.4 Table from Figure 6.3 with consistent formatting

The one exception to this Spartan view of spreadsheets is labels. All columns should have a clear and correct label at the top of the column and only at the top of the column (relating to the one table per spreadsheet rule). The label names should be short and you should abbreviate wherever necessary. The meaning of each abbreviation and further descriptions of the variables should occur externally to the spreadsheet in a data dictionary (see Chapter 4). You can also use a separate column for brief comments or an external file to document any notes about the data that do not fit into the spreadsheet formatting.

The final best practice for spreadsheets is to collapse smaller related tables into one large inclusive table whenever possible. This makes calculation easier over the entire dataset because the entire dataset is in one place with consistent formatting. Figure 6.5 shows how the smaller tables in Figure 6.3 can be collapsed into one larger table. Collapsing tables may require the repetition of some information, but the calculation benefits of one large table usually trump any space efficiency of smaller tables.

	A	B	C	D
1	Date	Species	Site	Count
2	20-Mar	Eagle	Eastbrook Park	0
3	20-Mar	Eagle	Lake Park	0
4	20-Mar	Eagle	Waterside Park	0
5	20-Mar	Gull	Eastbrook Park	2
6	20-Mar	Gull	Lake Park	12
7	20-Mar	Gull	Waterside Park	21
8	20-Mar	Mallard	Eastbrook Park	5
9	20-Mar	Mallard	Lake Park	2
10	20-Mar	Mallard	Waterside Park	4
11	20-Mar	Robin	Eastbrook Park	3
12	20-Mar	Robin	Lake Park	4
13	20-Mar	Robin	Waterside Park	0
14	21-Mar	Eagle	Eastbrook Park	0
15	21-Mar	Eagle	Lake Park	1
16	21-Mar	Eagle	Waterside Park	0
17	21-Mar	Gull	Eastbrook Park	1
18	21-Mar	Gull	Lake Park	15
19	21-Mar	Gull	Waterside Park	18
20	21-Mar	Mallard	Eastbrook Park	2
21	21-Mar	Mallard	Lake Park	2
22	21-Mar	Mallard	Waterside Park	3
23	21-Mar	Robin	Eastbrook Park	6
24	21-Mar	Robin	Lake Park	2
25	21-Mar	Robin	Waterside Park	3

Figure 6.5 Data from Figure 6.3 condensed into one table

6.3 MANAGING YOUR RESEARCH CODE

A great deal of research does not take place in the laboratory but occurs instead on the computer. Whether you are a theoretical physicist who works solely on the computer or a field biologist who creates one custom script for an analysis, your code is an important research product that requires care. This care is slightly different than managing data, coming with its own best practices and a unique infrastructure for versioning and sharing.

6.3.1 Coding best practices

Most research scientists are not trained programmers, so the process of coding is often one of the many skills picked up as a part of the research process. Compounding this challenge is the vast amount of code that gets passed between researchers in a laboratory, meaning researchers often deal with code that they did not write themselves. The end result is that there is a lot of research code out in the world that is unintelligible and irreproducible (and it's never fun to be on the receiving end of such code when you just want to do your research). There is a better way to deal with code in the laboratory and you don't need to be a programmer to get there (Wilson *et al.* 2014).

The first best practice of coding is that you should think about writing code that works for people. This means documenting your code, using well-named variables, and making it easy to follow the flow of the program. Great code looks nice, with consistent formatting and conventions. Great code also self-documents on what the program is doing, giving an overview of the "what" and "why" instead of describing the specific mechanics of each line of code. As with any documentation and formatting, you want someone who is unfamiliar with your code to be able to pick it up and use it without asking you for more information.

In terms of writing code, it's best to build code in small chunks and make sure each part works before creating the next bit of code. This will prevent the nightmare scenario of trying to fix all parts of your code at the same time – an impossible task if you cannot isolate sections as either correct or not. Realize that part of the process of writing code is fixing bugs, so build in debugging systems as you go. The simplest debugging systems print out variable values at key points in the program and more complex debugging leverages existing tools in your programming language. You may also be able to use your programming interface's debugger, a good one of which will walk you through the program step-by-step and allow you to examine variable values at any point in time. Finally, you should worry about making your code run properly before trying to make it run more efficiently. It's much easier to optimize everything when you have a working version.

All of these best practices and more are described more fully in an article by Wilson *et al.* (Wilson *et al.* 2014), which I recommend reading if you write a lot of code for research. Writing reusable code is not challenging, but takes patience in debugging and a little extra work to ensure that your code is reproducible and usable should you pass it on to someone else.

6.3.2 Version control

As coding becomes a more prominent part of many disciplines, we've seen the emergence of a number of useful tools for the creation of software. One that is particularly useful for scientific code is the version control system. At its most basic, a version control system allows you to track changes to your code over time and revert to earlier versions. This has useful applications for scientific analysis, among other things, as the version control system creates a complete record of your scientific processes. Version control systems also make changing your code easier, as you can maintain a working master copy of the code all while working on edits to a second copy, both managed by the system. Finally, modern version control systems facilitate collaborative coding, making the process of creating code for the laboratory a little bit easier.

If you maintain lots of research code, make frequent edits, or share code between your co-workers, I highly recommend adopting a version control system. It can make managing your code significantly easier. This section examines the version control system Git (www.git-scm.com), which is currently one of the most popular systems for version control and is commonly used for managing scientific code.

The fundamentals of version control

Version control systems for code are more complex than the simple file name versioning covered in Chapter 5. The main differences are that a version control system tracks many documents and only records the differences between versions of each document instead of saving each version as an independent file. Not only does tracking differences save storage space but it also directly identifies the changes from one version to the next, which is a useful feature. And of course, all version control systems are designed to let you revert to an earlier version as necessary.

The foundations of Git lie in the code "repository", which is the master copy of all the code for a project; separate projects usually have separate repositories. As you make changes to the code on your local computer, you periodically "commit" those changes to the master repository. Only through the act of committing does Git create new versions of any modified files and record changes from the previous versions. The distinction between committed and uncommitted changes allows you to write and debug code and only change the master copy of the code at important points, for example whenever the code runs correctly.

Code repositories can exist both on your local computer and remotely, and researchers commonly use GitHub (www.github.com) to host their repositories in the cloud. The benefits of using a remote repository are seen when multiple people work on the same code at the same time. In this case, individual users commit to local repositories and "push" changes to the shared remote repository (and alternatively "pull" changes from the central repository to update their local copy). Since Git generally handles the code integration, each user's main concern is making functional changes to the code.

Git's versatility in tracking edits and flexibility in handling many-user projects makes it a natural choice for managing research code. But even if you don't take advantage of Git's collaboration support, it's still a useful system for keeping track of changes in your own research code.

Using Git – the basics

If you are just starting with Git and don't regularly use the command line interface, I recommend using Git's graphical user interface (GUI). This interface is easier to start with than the command line because it provides you with a structured framework for interacting with Git. This section shows the Git GUI for Windows in order to help you better understand how Git works.

> **Learning Git on the command line**
>
> For those wanting to use Git on the command line, GitHub offers a great tutorial at http://try.github.io (GitHub 2014). This tutorial walks you through the basic commands of Git and lets you try things out on a test installation. The mechanics on the command line are equivalent to the processes you execute via the graphical user interface (stage, commit, etc.) but with written commands.
>
> In addition to learning Git by practicing commands, I also recommend writing up your own notes on commonly used commands. That way, you can record any useful information and express it in a way that makes sense to you. Git uses a lot of special terminology and you should use any tool that helps you familiarize yourself with the terms and variants.

No matter if you use the command line or user interface of Git, the first thing you do when using Git is to create a repository. The repository is associated with a group of files, tracks any changes made to them, and maintains a master version of them. While Git will automatically monitor these files, it won't officially record any changes until you commit them to the repository. Saving changes to the master files in the repository requires two steps: staging and committing. Committing allows Git to officially record changes and add them to the master copy of the code. Staging identifies the files you want to commit, letting you commit one or more files at a time without committing every file that has been changed. This is a useful feature for when you are creating and debugging code. Between staging and committing, you should write a description of the changes you are committing; this is your "commit message". Commit messages help you, the user, keep track of changes and are something you should get in the habit of creating when committing. Follow the whole committing procedure any time you want to commit changes to your repository: edit files in the file's native interface, open Git, stage files, write a commit message, and commit files.

You can visualize the commit process in the Git GUI shown in Figure 6.6. The interface is divided into four panels: unstaged and staged changes in the upper and lower left respectively, a panel to view changes to any file in the upper right, and a place for writing "commit messages" in the lower right. So we can see that Git keeps track of which files are different than the master version (upper left), what those differences are (upper right), what files are staged and ready to commit (lower left), and what information you are recording about the changes (lower right). This represents some of the most basic functionality of Git, no matter if you are using the command line or the Git GUI.

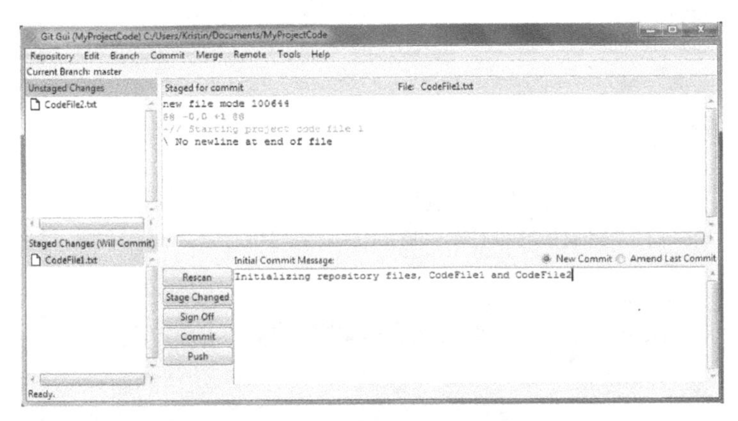

Figure 6.6 The Git graphical user interface, ready to commit "CodeFile1.txt"

Once you start committing files and changes to your repository, Git can show you the history of the changes to any one file. In the Git GUI, this is represented as the History window shown in Figure 6.7. The History window provides a lot of information, including a list of commits, commit messages, and changes to your file. You can also use this window to revert to an earlier version of your code or undo reverts by jumping forward in the process of edits. This is the real power of the version control system.

Git is a powerful system for keeping track of changes to code and, whether you use the Git GUI or the command line, this tool can be a boon to your research workflow. To integrate Git into your research, start with basic commits and slowly build your Git vocabulary until you have a tool that suits your coding needs. There are plenty of Git resources available to help you on this journey (Chacon 2009; Loeliger and McCullough 2012; Lawson 2013; Mathôt 2014) and it's worth checking them out if you want to get serious about Git.

6.3.3 Code sharing

Like data sharing, sharing and getting credit for your research code is slowly becoming a more common practice in research. There are many reasons for this (Stodden 2010; LeVeque *et al.* 2012), including increasing reproducibility by being more transparent about analysis methods, seeing code as a valuable research product, making your research higher impact, and advancing science by allowing others to view and use the latest computational methods. Indeed, some journals (Stodden *et al.* 2013) are now requiring that code be shared in conjunction with article publication and there is even a whole website, RunMyCode (www.runmycode.org), dedicated to testing out others' scientific code. I encourage you to consider the value of your research code and how it might be shared to improve science.

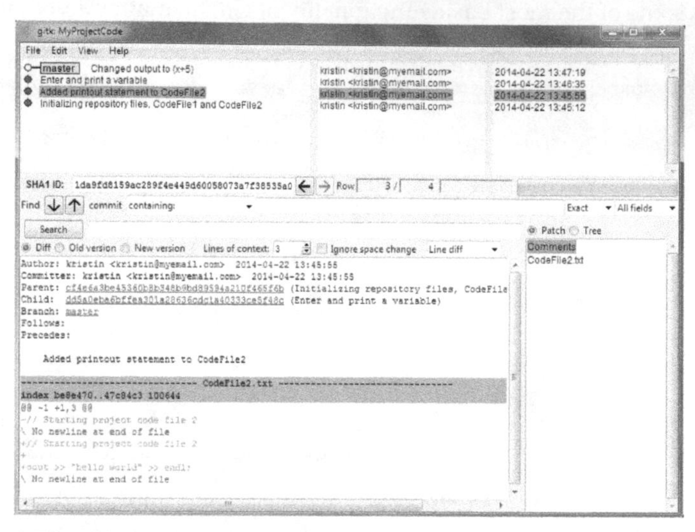

Figure 6.7 The GIT graphical user interface History window

Sharing code has a different infrastructure than sharing data, as it can build upon existing resources for open source software. The two foremost resources for openly sharing research code, GitHub (https://github.com) and Google Code (https://code.google.com), are two platforms based in the open source community but commonly used for disseminating research code. GitHub, in particular, is a good option if you are using Git for version control as it seamlessly integrates into versioning and acts as a remote repository. GitHub and Google Code are the best options for code sharing because of added functionality, but you can use some data repositories, such as figshare (www.figshare.com), to share research code. There are also "traditional" scholarly ways to share code and the Software Sustainability Institute provides a nice list (Chue Hong 2013) of journals that publish research code in a range of scientific disciplines.

As with data, it's important to remember the extras when sharing code. This means you should add general documentation and note any publications that rely on this code. If you are sharing code, you should also think about sharing the corresponding data for the sake of reproducibility. This is particularly important if you are doing a simulation that requires seeding, for example with a set of random numbers; you should share the seed values so that users can recreate your exact results. Sharing code is extremely valuable so long as you put it into context of how it is used in research.

6.4 CHAPTER SUMMARY

This chapter covers many aspects of data management that make analysis easier. Such practices include maintaining copies of the raw data, the final processed data, and any data at important intermediate analysis steps; this allows you to easily re-analyze data should the need arise. You should also document your analysis process for the same reason.

Prior to doing your analysis, you should perform some basic quality control on your data. This means checking your data for errors, making your data more consistent, using appropriate null values, and avoiding compound variables. If you use spreadsheets, your data should be collapsed into as few tables as possible, each with its own spreadsheet. Spreadsheets should have a label at the top of each column and no extraneous formatting.

If you work with code, you should make it as human readable as possible with good variable names and appropriate documentation. Build in debugging systems as you code and expect this to be a significant part of the coding process. If you have a lot of code, a version control system such as Git will help you manage changes to your code over time. Finally, consider sharing your research code to improve your research impact; you can leverage existing open source software tools to do this.

7

MANAGING SENSITIVE DATA

In 2009, the University of North Carolina discovered a data breach in the lab of breast cancer researcher Bonnie Yankaskas (Barber 2011; Ferreri 2011; Kolowich 2011). The affected server contained information on 180,000 women – including names, addresses, birth dates, and social security numbers – all potentially now in the hands of the hackers. The university moved to deal with the incident by alerting victims of the hacking and by trying to fire Yankaskas. Yankaskas, in turn, contended that she is not an information technology expert and should not be punished for something she was not responsible for – namely, directly maintaining a secure server. The argument between university and researcher lasted for two years. In the end, Yankaskas was forced to retire.

The Yankaskas dispute is an enlightening one for researchers working with sensitive data because it shows that, while you may not have expertise in data security, you are still responsible for maintaining data securely. In Yankaskas' case, that means she should have hired competent IT personnel, dedicated sufficient funding for data security, and followed best practices such as avoiding the storage of social security numbers. In turn, the university's obligation was to provide sufficient IT infrastructure. Too many players in this drama assumed sufficiency when the system was actually a collection of shortcuts, resulting in risky storage practices.

The moral of this story is that, while I agree with Yankaskas that researchers should not need to be security experts, you need to know the basics of security and work closely with your local security experts to keep your sensitive data safe. Using that philosophy, this chapter covers a general outline of practices to help make your data more secure. Beyond that, I strongly urge you to find local experts and resources to ensure your security practices are sufficient. Because if the Yankaskas story teaches us nothing else, it is that you must be proactive about data security and work with people who are experts in the field.

7.1 TYPES OF SENSITIVE DATA

When working with sensitive data, there are three general recommendations: determine if your data is sensitive, don't collect sensitive data if you don't have to, and make a clear security plan with help from security experts and review it frequently (see Section 7.2.6). It makes sense to start a chapter on sensitive data with the first recommendation, by determining what data is sensitive and identifying any corresponding requirements, so that we have a framework for implementing the security practices covered later in the chapter. So in this section we will examine sensitive information protected by national law, data covered by research ethics, and other types of data often categorized as sensitive.

7.1.1 National data privacy laws

Research data containing personally identifiable information is frequently subject to national law governing how that data should be handled and retained. Unfortunately, these laws vary from country to country so there is not one magical strategy for bringing your data practices into compliance with any country's law. This means that, while we will examine a few of these laws here (see Table 7.1 for an overview), you should consult with your local security expert on the specific requirements of the laws that apply to your data.

Table 7.1 National laws and international directives on data privacy

Law/Convention	Domain	Description
Data Protection Directive	European Union	Regulates personal data and directs the creation of similar laws by member nations
Data Protection Act 1998	UK	Broadly regulates storage and use of personal data in compliance with the Data Protection Directive
Bundesdatenschutzgesetz	Germany	Broadly regulates storage and use of personal data in compliance with the Data Protection Directive
The Privacy Amendment	Australia	Broadly regulates storage and use of personal data
HIPAA	USA	Regulates storage and use of personal health information
FISMA	USA	Regulates storage of data for parties working as contractors for the federal government

Some of the more consistent data laws come from the European Union, which adopted the Data Protection Directive in 1995. In addition to laying out requirements for working with any type of personally identifiable information, this directive requires member nations to bring their national laws into compliance with the directive. Therefore there are many European national laws that regulate data containing personally identifiable information, such as the Data Protection Act 1998 in the UK and the Federal Data Protection Act, Bundesdatenschutzgesetz, in Germany. In general, these laws dictate that personally identifiable data must be:

- Processed fairly and lawfully
- Collected for specified and legitimate purposes
- Adequate and relevant to the purpose for which the data is collected
- Accurate and kept up to date
- Kept in a form which permits identification for no longer than necessary (European Parliament 1995)

These laws also dictate who can access the data, how data must be stored, and processes for consent. Note that in 2012 the European Union began discussion on an updated version of the Data Protection Directive, which will affect all of the EU member nations.

Researchers in Australia must work within The Privacy Amendment (Enhancing Privacy Protection) Act 2012 from Australia, which is similar to EU regulations. This law applies to anyone using personal data in Australia as well as companies operating in the Australian market. The amendment contains 13 privacy principles:

- Consideration of personal information privacy
- Anonymity and pseudonymity
- Collection of solicited personal information
- Dealing with unsolicited personal information
- Notification of the collection of personal information
- Use or disclosure of personal information
- Direct marketing
- Cross-border disclosure of personal information
- Adoption, use or disclosure of government related identifiers
- Quality of personal information
- Security of personal information
- Access to personal information
- Correction of personal information (Office of the Australian Information Commissioner 2014)

These points exhibit broad coverage of how to handle and use data containing personal information.

The United States is different from other countries in this area, as there is not one law that broadly covers all personally identifiable information. Instead, the US has a patchwork of more focused regulation governing particular types of information, such as student records. The laws most likely to affect scientific researchers are the Federal Information Security Management Act (FISMA), which mainly applies to government contractors, and the Health Insurance Portability and Accountability Act (HIPAA), which broadly protects human health information. FISMA covers those working with government data and requires government-level security with corresponding monitoring and reporting. If you fall under FISMA, get competent technology support to help with compliance. The other relevant privacy law, HIPAA, dictates how human health information can be stored and who is allowed access to it, in addition to defining ways to anonymize data so that it can be safely handled by non-authorized users (see Section 7.3.3 for information on HIPAA's "Safe Harbor" rules for anonymization). Do note that HIPAA only applies to health information; publicly available or corporation-held personally identifiable information unrelated to someone's health is not covered under this statute.

Due to the variety of requirements in data privacy laws, it is best to be aware of the individual laws and policies that apply when your data contains personal information. While such requirements add an extra layer to your security strategy, they also give you guidelines for how you must handle certain types of sensitive information.

7.1.2 Ethics and sensitive data

Even if you have no law or policy specifically governing the handling and retention of your data, research ethics require you to securely maintain data containing things like human subject information. A researcher entrusted with personal data has an obligation to the research subject to protect the subject's information. We therefore label human subject data as sensitive for ethical reasons.

Research ethics obviously entail more than keeping data secure. For example, the Framework for Research Ethics from the UK's Economics and Social Research Council lays out the following ethics requirements:

> 1. Research should be designed, reviewed and undertaken to ensure integrity, quality and transparency.
>
> 2. Research staff and participants must normally be informed fully about the purpose, methods and intended possible uses of the research, what their participation in the research entails and what risks, if any, are involved. Some variation is allowed in very specific research contexts for which detailed guidance is provided in Section 2.
>
> 3. The confidentiality of information supplied by research participants and the anonymity of respondents must be respected.
>
> 4. Research participants must take part voluntarily, free from any coercion.
>
> 5. Harm to research participants and researchers must be avoided in all instances.
>
> 6. The independence of research must be clear, and any conflicts of interest or partiality must be explicit. (Economics and Social Research Council 2012)

Basically, these requirements state that researchers must keep data confidential, obtain informed consent, avoid harm to research subjects, and avoid conflicts of interest. Researchers doing human subject research in the United States are similarly governed by the Common Rule. All of these ethical requirements are usually enforced via approval by a local Internal Review Board (IRB) or Ethics Review Board (ERB). If you conduct human subject research, you should become familiar with your local board's policies and approval process.

Putting all this in terms of data management, there are two main concerns with this type of data. The first is to obviously keep sensitive data secure. The second is that, if you anticipate data sharing and reuse, informed consent should not preclude potential data reuse. This can be achieved via open consent (also called broad or enduring consent), which covers how data might be reused instead of dictating that data will be destroyed at the end of a study (Lunshof *et al.* 2008). While open consent cannot be as specific about future uses of the data as traditional consent, there are indications that it does not diminish participation and, in fact, participants like the fact that their contributions can continue to advance science (Corti 2014). So consider using open consent if your sensitive data has reuse value or you are subject to data sharing requirements.

7.1.3 Other data categorized as sensitive

So far we've examined sensitive data in the form of personally identifiable information, but there are other types of data that fall under the "sensitive" label. One type is information that could cause harm if publicly released. A lot of government information falls into this category, though occasionally this category pops up in research such as for the 2012 moratorium on H5N1 avian influenza research and publication (Enserink and Malakoff 2012; Keim 2012). It's usually obvious if you are working with the government using confidential data, but ethics and review boards can provide guidance on other types of information that should not be publicly released.

Another type of data often labeled as sensitive is data leading to patents and other intellectual property. This data is not sensitive in the personal or legal sense, but sensitive in that it is proprietary and should not be publicly released. Corporations, in particular, are likely to view much of their data as sensitive and take steps to protect information from theft and public release, especially during the research and development stage. While sensitive intellectual property data does not legally require restricted access, you are likely to find several strategies in this chapter that you can use to prevent early disclosure. See Chapter 10 for more information on data and intellectual property.

Finally, even general data benefits from basic security practices, especially the basic computer security practices outlined in Section 7.2.1. This particularly helps the many researchers concerned about being scooped. By following standard security practices, you can prevent other researchers from accessing your data prior to publication. So even if your data is not legally or ethically required to be treated as sensitive, consider using the practices covered in this chapter to prevent unwanted access.

7.2 KEEPING DATA SECURE

The easiest way to deal with data security is by not collecting sensitive data in the first place. This allows you to avoid all the effort required to keep data secure.

Unfortunately, it's not always possible to do research without collecting sensitive information. For this reason we will now examine some of the practical aspects of data security.

Keeping data secure involves a lot of small practices that add up to a secure storage and working environment. These practices include having safe computer practices, limiting access to the data, encrypting your data, properly disposing of data, hiring competent people, and training everyone in security procedures. Basically, you want to maintain a safe storage environment and not remove sensitive data from it. In addition to all of these practices, you should consult with your local security experts to make sure you are handling your data properly. When it comes to data security, it is always best to display a little extra caution.

7.2.1 Basic computer security

A good portion of keeping data secure is simply following good computer practices. With an increasing number of hackers, spammers, phishers, and general ne'er-do-wells trying to gain access to vulnerable computers, the first step in data security is actually computer security. You must keep your computer secure in order to keep the computer's contents secure.

Software

The first good computer practice is to keep your software up to date. In particular, you should regularly update operating systems, browsers and anything else that accesses the internet, and anti-virus and other software that protects your computer from the outside world. Software must be frequently patched in order to prevent vulnerability to the computer bugs, viruses, and security holes that arise with surprising frequency. You leave yourself vulnerable to attack by not regularly updating software. The good news is that operating systems, browsers, and many other types of software can be configured to update automatically, making it easy to keep your software patched.

Beyond keeping your software up to date, you should employ certain types of software to protect your computer, namely: anti-virus software, anti-malware software, and a firewall. This suite of software will help protect you from threats and help you find and destroy malevolent programs that are already on your computer. While firewalls run constantly when they are turned on, you should set anti-virus/malware software to automatically perform routine scans.

Practice safe usage

You should also practice safe behavior when using your computer. For example, stick to known sites when browsing the web. This also applies to downloading, opening, and running files and installing new software. If you do not know and trust the source, do not click on the link, download the file, or install the software. Spammers, in particular, have grown quite sophisticated in recent years, so err on the side of caution. If you have any doubts, consult a local IT expert.

To take this one step further, there are ways to separate yourself from some of these issues. One way is to access and browse the internet on one computer and reserve a separate computer for your secure data. You can also limit your use to a non-administrative account on a computer. Such accounts allow you to use the computer but not to install new software, creating a security barrier against potential threats. If you are browsing the web using a public wireless network, also consider using a virtual private network (VPN) to protect against notoriously unsecure public Wi-Fi. Each of these strategies helps keep your computer more secure.

Passwords

The final general recommendation is to use strong passwords. Strong passwords make it hard to access whatever is behind the password because the password cannot be guessed. Strong passwords also stand up well to brute-force attacks on your computer, wherein an outside computer tries a huge number of password variants until it stumbles upon the correct one. A strong password exhibits several characteristics:

- Uses a mixture of letters (both upper- and lowercase), numbers and characters
- Is at least eight characters long
- Is non-obvious
- Is not a dictionary word
- Is not a proper name

Dictionary words, names, and obvious passwords make poor passwords because they are easy to guess, even if they are longer than eight characters. While dictionary words are easy to define, obvious passwords are a little more nebulous. So let's look at five obvious, yet commonly used, passwords:

- 12345
- qwerty
- password
- monkey
- abc123

Needless to say, do not use any password on this list. Other obvious passwords include things like your child's or pet's name, your birth or marriage date, and your user name. Generally, you should avoid using passwords that someone could guess, given a little knowledge about you.

In lieu of an obvious password, one strategy to get a strong password is to start with a couple of words or a short phrase you will remember and add variations. This could be capitalizing letters, adding a couple of numbers, and/or throwing in a random character. For example, the phrase "I love research" could become the password "iLOVEr3search!" and "the ants go marching two by two" could be "@ntsMARCH2x2". Alternatively, you can start with a longer phrase and abbreviate it. For example, the R.E.M. lyric "it's the end of the world as we know it" could be a password as "iTeOtWaWkI". These strategies take something you will remember, use a little creativity, and transform it into a strong password. A third alternative is to make up a nonsense word that is pronounceable, such as "valistraten" or "encroptisment". Being pronounceable makes the password more memorable but such passwords are still non-obvious. Ultimately, you should have a password that is memorable but complex enough to stand up to password guessing and brute-force attacks, the most common hacks.

Once you create a strong password, you should not use it for any other login credential. This is because if one site is hacked, the hacker now has access to your account on all of the other sites for which you used the same password. Get in the habit of changing your password any time a system gets hacked, breached, or infected with a virus.

To help with the profusion of passwords, you can employ a password manager. Password managers are software that remembers your passwords for you, allowing you to use stronger passwords than ones that need easy remembering. Such technology is a good idea for general passwords but not for passwords protecting data requiring a high level of security, such as that containing personally identifiable information. It's best to have direct control over passwords for highly sensitive content, but for the rest, a password manager is a handy tool.

Finally, you should never, ever share your passwords with anyone else. Ever. Even people without devious intentions can cause harm when given access and, since it is using your login credentials, you are ultimately responsible for the consequences. It is better to appear rude by not sharing a password than to risk harm by allowing someone access to your precious data. So create strong passwords and keep them to yourself.

7.2.2 Access

In essentials, protecting sensitive data means preventing unauthorized access. This can be achieved in many ways, ranging from firewalls to encryption, but there are some fundamental access practices you can do both physically and digitally to keep data secure.

Maintain a secure storage environment

The first rule of data security is that data should always stay in its secure, controlled-access environment. Moving sensitive data out of this environment enables outsiders to more easily access the data, resulting in a data breach. Unfortunately, the side effect of keeping data in a secure environment means that it is sometimes difficult to work with; you must go to the data instead of working wherever is convenient at that moment.

A secure environment entails both physical and electronic methods of preventing access. The most basic way to deny physical access is to put data behind a locked door and/or in a cabinet or drawer with a lock. This works to protect computers and other digital storage media containing sensitive data. On the electronic side, passwords protect against unauthorized digital access (see Section 7.2.1 for more information on passwords). You should also set up a secure computer/server to store your data. For added security, disconnect the system from the internet. Remember that multiple levels of protection, such as locking a password-protected device in a drawer in a locked office, make it harder for an unauthorized individual to get the data.

Maintaining a secure environment also entails avoiding storage options that are easily accessible. Two particularly vulnerable places for sensitive data are email and cloud storage, though you should not place sensitive data on the internet in general. Mobile devices are also very data porous and are thus not recommended. Data placed onto these platforms becomes data that is no longer under your direct control, either because the data lives on someone else's servers or because it becomes easier to access data in an unsecure environment. If you absolutely must use these platforms, de-identify the data before uploading it.

A lot of secure data is lost in the in-between places – for example, when someone downloads a secure dataset to work on an unencrypted laptop, which is then stolen – so you must be extra vigilant about keeping secure data only in secure places. It obviously means more work, but the alternative is to risk your career and your employer's reputation.

When you have to move data

While the first principle of access is to keep sensitive data in its secure environment, there are a few cases where it is difficult to avoid moving data. Researchers often need access from a secondary location, for example, if they collect sensitive data in one location but store it in a geographically different location or when they share data with an offsite collaborator. In these cases, you want to plan very carefully ahead of time how you will maintain data security during every part of the process. This is especially important for data that is legally regarded as sensitive.

The first part of planning is to ask if it's absolutely necessary to collect the sensitive portions of the data. Often, it's much easier to not collect sensitive data such as social security numbers than to deal with added security concerns. If you must collect sensitive data, determine if it's possible to bring the researcher to the data instead of the other way around. You can do this by physically coming to the data or by performing remote computation on the data so that you do not make a local copy – note that this is something you should get help from an expert to set up. Finally, if you must move the data to a secondary location, de-identify it before transfer (see Section 7.3) or build two secure environments and transfer the data in an encrypted form. Again, ask for help as there can be complications in moving even encrypted data. For example, security agents have been known to confiscate computers and other devices crossing international borders. While this is an extreme example, you are generally more likely to lose devices in transit than in a locked down environment, so you want to plan ahead.

Controlling access

Blocking outsiders is a big part of controlling access but access also entails keeping track of those allowed to use the data. In practice this means monitoring who has what permissions over what data and documenting when people access that data. The latter is done using access logs. For physical data, researchers often use a sign-out sheet to record who borrowed the data, what they borrowed and when they borrowed it, and when the researcher returned the data. For digital data, it's best to set up logs that automatically track what people work with and when. Access logs can make all the difference if a problem arises with the data, such as unauthorized access or misconduct within the laboratory. For example, serial data fabricator Jatinder Ahluwalia first came under suspicion for odd test results and citing experiments that did not have corresponding (and legally required) radioactivity, animal housing, and disposal records (Oransky 2011). Record keeping takes extra work, but it's necessary for highly sensitive data.

The other part of access control is providing researchers with the proper permissions to use the data. This usually isn't a problem to set up but often becomes a problem later. That is because, while researchers remember to give access rights to someone starting to work in the laboratory, they often forget to revoke access after someone stops working in the lab. This is a huge security hole and there are, unfortunately, stories of ex-employees making mischief with content to which they should no longer have access. Properly cutting off access includes taking back keys, deleting user accounts, taking back physical data and media, and any other steps necessary to remove access. Make it a regular part of the leaving procedure to check people's access to secure data.

7.2.3 Encryption

A chapter on keeping sensitive data secure would not be complete without a discussion on encryption. Just as the Germans used the Enigma machine during the Second World War to encode messages, so too can we encode sensitive data files – and even whole devices – so that content is unintelligible but still retrievable. Encoding and encryption technology has obviously changed over the years, but the general principles are still the same. Namely, those with the proper key or passphrase can access the data and those without the key cannot. And just like with the Enigma machine, we also need to worry about outsiders cracking our encryption if we do not use the best practices and technologies.

Encryption algorithms

At its most basic, encryption programs take digital files and, using a user-supplied "key" and a particular algorithm, transform the bits in the file into a different collection of bits. The algorithm cannot decrypt the data – that is, transform the encrypted file back into the original document – without using the proper key. Encryption works such that if a potential hacker knows the encryption algorithm but not the key, then the hacker cannot gain access to the data. Note that this premise does not hold true for all encryption algorithms; computing has advanced such that algorithms once considered secure can now be easily broken with modern computational resources.

You have several options in choosing encryption algorithms, and thus software, though the best option is usually the one offered by your local technical support. Using locally supported software saves you from vetting software and likely provides you with assistance should questions arise. In the absence of a supported option, look for something that uses the popular Advanced Encryption Standard (AES) algorithm. Alternatively, you can use Pretty Good Privacy (PGP) software, which is also available in open source as OpenPGP.

Encryption keys

Besides the algorithm, the other important part of encryption is the key. Encryption is useful principally because you trade the big task of keeping a whole file secret for the smaller task of keeping the key secret. Needless to say, that means you should not share encryption keys except with the people who need access to the data.

Modern encryption programs rely on several different types of keys, depending on the encryption software. One common type is the password or passphrase. Passphrases are like passwords but longer, typically at least 20 characters (Smith 2013), and can include spaces in addition to other types of punctuation. The principles for creating a passphrase are similar to those for creating a password (see Section 7.2.1 for guidance on creating strong passwords) – namely, that you should use many types of characters to come up with something that is difficult to guess. Remember that you must keep your passphrase secret, so choose something you can easily remember instead of something you need to write down to remember. For added security, consider changing your passphrase every two years (Smith 2013).

Another popular type of key is the public/private key pair. This key pair operates by having a public key that can be used for encryption and a secret key, the "private key," that can retrieve information encrypted by the public key. Thus, anyone with the public key can encrypt information but only someone with the proper private key can decrypt it. This is useful for sending encrypted information between two people. As with passphrases, you should always keep private keys secret. The public/private key combination is also used in reverse to form digital signatures; the private key signs a document and the public key validates that the document was signed by the private key's owner. The ability of these two keys to work in tandem to encrypt and decrypt information is based on the fact that the keys are mathematically related but it is incredibly difficult to compute the private key from the public key. The reliance on private/public key pairs depends on the software, but this technology is particularly useful when you want to share encrypted information with another person.

Other encryption considerations

In conjunction with choosing encryption software, you also have the choice of encrypting single files or the whole storage device. Single-file encryption is a good option if you only have a few sensitive files. Beyond that, consider encrypting your whole device. The benefit of doing this is that whole disk encryption covers all files, including temporary files or overlooked sensitive documents. Additionally, disk encryption is preferred if you place sensitive data on removal storage media, such as an external hard drive. For added security, you can also combine single file encryption and whole disk encryption.

The final thing to note about encryption is that, while encryption makes it easier to store sensitive data, it doesn't mean that you can stop worrying about data security. In fact, relying overmuch on encryption can open your data up to security risks, mainly because you have to decrypt the data at some point if you want to work on it. At this point, it's possible for others to access your data if you do not maintain a safe computer environment. The other major risk is that it is easier to move data that is encrypted to offsite locations. If you then decrypt the data, suddenly you have lots of sensitive data in an offsite, and usually insecure, environment. Therefore, you should stay vigilant about data security even when your files stay encrypted most of the time.

Encrypted cloud storage

One option you might encounter when considering encryption technologies is encrypted cloud storage. When done right, encrypted cloud storage is a good option for storing data but it can still run up against cloud storage's main problem: the easy propagation of datasets. So encrypted cloud storage is useful for data that needs secure storage but not for sensitive data, unless that data is only accessed from secure computers.

To pick a good encrypted cloud storage provider, look for services that encrypt data locally before sending the data to the cloud. This prevents anyone from intercepting the data over the internet and makes sure that the service provider cannot read any of the files on their servers. There is little point in using encrypted cloud storage unless your data is encrypted along the whole path from your computer to the cloud and back.

Additionally, it's always a good idea to maintain a backup copy of the data. If we learned anything from the research data-focused cloud service provider Dedoose's data loss in 2014 (Kolowich 2014) or the previous year's Cryptolocker virus that encrypted computers for ransom (Ward 2014), it's that you always want a backup copy of your files.

7.2.4 Destroying data

Data security requires proper storage habits from the beginning to the end of the life of a dataset which, in practice, means destroying data after it is no longer needed. Only by destroying data can you ensure that it is not at risk for loss after that point. Additionally, destroying data means that you no longer need to devote resources to maintaining a secure storage environment for that data. It is best practice to destroy sensitive data after it is no longer needed or required to be retained.

Destroying sensitive data requires more work than simply deleting the file from your computer. When you delete a digital file from your computer, the file stays on your hard drive and only the reference to the file is removed. This is why it is often possible to recover data from a hard drive after a crash; the files are not truly destroyed even if the operating system does not recognize that they are still there. To truly delete something from a hard drive requires a specialized type of software; consult with your local technical support to see what data deletion software or support is available to you.

While hard drives can achieve true deletion via software, other types of media require different methods for data destruction (Corti 2014). The mechanics of storage on flash media, such as thumb drives, entail destroying the entire device to truly destroy data. It is also difficult to achieve true deletion on solid state devices (Johnston 2011; Wei *et al*. 2011), so ask for help if you need to destroy data on this type of media. Like flash media, CDs and DVDs must be physically destroyed, with breaking or shredding being a good option for this. It is a good idea to consult with your local information technology specialists for resources and assistance with destroying any of these types of media.

It is much easier to delete datasets on paper, as destroying the paper destroys the dataset. The most common way to destroy paper-based information is shredding. For extra-secure destruction, paper can be cross-cut into small pieces of confetti. Don't forget to destroy consent forms, surveys, and other paper-based sensitive information, as needed, in addition to sensitive data.

Finally, note that there is software for finding sensitive data, such as social security numbers and bank account numbers, on your computer. This is especially useful for making sure that data has not spread to unrestricted areas and to gather up all sensitive data for destruction. If you think you have a problem with data spreading to uncontrolled locations, look into "PII scanning software" (personally identifiable information scanning software) such as Identity Finder or OpenDLP.

7.2.5 Personnel

Researchers with access to sensitive data add another vector of security risk. As they have access, it is much easier for them to accidentally or purposefully alter data. For example, Jatinder Ahluwalia, mentioned earlier in this chapter, not only fabricated his own data but also sabotaged colleagues' experiments so that they unknowingly published false data (Oransky 2010). People are also vulnerable to social engineering, wherein they are tricked or pressured into giving up login credentials. So just as you should never share passwords, you should also never log someone else into an account under your name. All this leads back to the fact that you should only provide access to sensitive data to people you trust.

This trust extends to technical support and is particularly important when you hire someone to manage security for you. Note that if you have a large or complicated amount of sensitive data to protect, you should be hiring someone to take care of your data security. Do not cut corners on cost or expertise here. The story of Dr. Bonnie Yankaskas at the beginning of the chapter demonstrates the dangers of not using qualified professional staff as IT support. Hiring competent support costs money, but comes with less risk for your data.

7.2.6 Training and keeping a security plan

The final part of keeping data secure requires that everyone who comes into contact with the data knows the security procedures. For this reason, it's recommended to codify your security practices in a written document. A written document helps everyone understand what is expected and provides a framework to refer to when there are questions. Plan to collectively review the security practices on an annual basis to ensure that everyone is following the procedure and to update practices as necessary.

In conjunction with reviewing the security plan, you should have regular training sessions for your entire research group. Training should occur when a new person enters the laboratory and on an annual basis thereafter. Regular review helps ensure that people are following best practices and that there is an opportunity for discussion if a particular procedure is not working well. Finally, if you have outside training opportunities, take advantage of them! It is better to have too much knowledge about data security than not enough.

7.2.7 Summarization of the dos and don'ts

This section covers quite a few recommendations for maintaining data securely. For easier reference, here are those recommendations summarized as a handy list of dos and don'ts:

- Do keep your operating system and software patched and up to date
- Do use anti-virus, a firewall, and anti-malware software
- Don't visit unknown sites on the internet
- Don't open suspicious attachments or click on suspicious links in emails
- Don't browse the web on the computer you use for sensitive data storage and analysis
- Do use strong passwords
- Don't repeat passwords
- Don't share your passwords with others
- Don't collect sensitive data unless you have to
- Do encrypt your sensitive data
- Don't move sensitive data outside of its secure storage environment
- Don't store sensitive data without using both physical and electronic safeguards
- Don't put sensitive data on the internet, in the cloud, or in email
- Do plan how you will safely move sensitive data before you collect data in a secondary location
- Don't move unencrypted, identified data
- Do use logs to monitor access to sensitive data
- Do cut off access when someone leaves the group
- Do destroy sensitive data once it is no longer needed

- Don't provide someone with access to sensitive data if you do not trust them
- Do hire competent technical support
- Do make a security plan and frequently review it with others

7.3 ANONYMIZING DATA

Up to this point, the focus of this chapter has been on how to manage and secure sensitive data. There is another option however, and that is to transform sensitive data into non-sensitive data. This process is called anonymization and it's used on data containing personally identifiable information. Not every dataset can be anonymized but, where possible, anonymization is often worth the effort because it means that you no longer have to treat your data as sensitive. This book provides an introduction to the topic of anonymization. If you decide to anonymize your data, plan to become more familiar with your preferred anonymization method.

There are lots of reasons to anonymize your data. The first is financial. Anonymized data does not need a special secure environment for storage. Additionally, if you lose an anonymized dataset, you are not liable for damages as for a sensitive dataset. One estimate puts notification of a data breach containing personal health information at $200 per person (El Emam 2013), a cost that can add up if a large sensitive dataset or database gets hacked. The second reason is that it is less of a hassle to work with anonymized data. Most data privacy laws no longer apply once you remove the personally identifiable information from a dataset, which means that you don't have to adjust your routine as you would to work with a sensitive dataset. Finally, anonymized data is shareable data. This is partly because data does not contain sensitive information that must be strictly controlled, but also because anonymization can allow for data reuse outside of the sensitive data's original purpose.

There are, of course, downsides to anonymization. One major downside is the fact that you lose some information through the anonymization process. For example, anonymization may reduce specificity in the case of a person's age or address or remove whole data points such as when a dataset contains a few easily identified individuals. The goal of good anonymization is to balance loss of information with the risk of identifying a particular person in the dataset, called "re-identification". The other downside is that performing good anonymization is not easy. There are many stories of poor anonymization, such as when AOL infamously released millions of poorly anonymized search queries, resulting in several people losing their jobs at the company (Barbaro and Zeller 2006; Electronic Privacy Information Center 2014). Still, by following the current standards for anonymization, you will significantly lower the risk that your anonymized data will be re-identified.

Another concern about anonymization is the conviction by a number of scholars that a dataset can never be truly anonymized (Narayanan 2014). Between advances in the technology of re-identification and the compilation of ever-larger datasets as part of "big data", it is difficult to predict how an anonymized dataset may be used or correlated with other data in the future. For example, it is now possible to re-identify someone from their genome (Check Hayden 2013), something that would not have been possible ten years ago. All this is to say that anonymization may not be guaranteed, though there is currently no consensus on this issue. What is known is that you should use the current best standards for anonymization. Best practices may eventually dictate not using anonymization at all but anonymization is currently a valid method for dealing with sensitive data.

7.3.1 Types of personally identifiable information

Before addressing the mechanics of anonymization, we must first examine the types of data requiring anonymization. Personally identifiable data comes in two varieties: direct identifiers and indirect identifiers. Direct identifiers are things that identify a person individually. They include:

- Name
- Address
- Telephone number
- Email address
- IP address
- Social Security Number, National Insurance number, or other national ID
- Driver's license number
- Medical record numbers
- Credit card number
- Photographs
- Voice recordings

Basically, any one item that can usually distinguish one individual from everyone else is a direct identifier.

Unlike direct identifiers, indirect identifiers do not disambiguate with one data point but instead can be used in combination to identify an individual. For example, security researcher Latanya Sweeney showed that 87% of Americans can be uniquely identified by their zip code, birth date, and gender (Sweeney 2000). Alone, none of these data points is a direct identifier but they can be used collectively to identify someone. Examples of indirect identifiers include (El Emam 2014):

- Gender
- Ethnicity
- Birth date
- Birthplace
- Location
- Marital status
- Number of children
- Years of schooling
- Total income
- Profession

- Criminal record
- Important dates (hospital admission, medical treatment, death, etc.)
- Disabilities

As indirect identifiers also provide the ability to identify individuals, it's important to also anonymize them in a dataset.

7.3.2 Masking data

When faced with personally identifiable information, you generally have two choices for how to anonymize your data: masking and de-identification (covered in Section 7.3.3). Masking is a blunt strategy involving the suppression of identifiable information. While masking serves the purpose of anonymization, it usually results in significant holes in a dataset, making the data less usable. Still, masking is worth discussing because it is an effective method for anonymization and its techniques are also useful for de-identification.

Masking involves removing or obscuring the identifiable information in a dataset. Three techniques exist to do this: suppression, randomization, and pseudonymization. These three techniques can be used individually or in combination to anonymize any dataset.

The first masking technique is suppression. Suppression involves removing, leaving blank, or setting to null all of the fields that contain personally identifiable information in a dataset. For example, suppression would entail entirely removing columns corresponding to patient name, medical record number, etc. in a spreadsheet containing medical records. While this anonymization technique is effective, it is not always useful. For example, there are cases such as software testing where you want to be working with a realistic dataset. Therefore, we must consider randomization and pseudonymization.

Randomization and pseudonymization involve replacing identifiable information instead of removing it. Randomization entails replacing identifiable content with random values. For example, you would indiscriminately replace first and last names with new names. This replacement is not consistent across the dataset, making it impossible to correlate related information in a randomized dataset. Pseudonymization solves this problem by replacing identifiable information with consistent pseudonyms. For example, a dataset might always replace the name "Gina Porter" with the name "Diana Stevenson" to anonymize the data but keep all of Gina's records together. Another option for replacement is hashing, which is discussed in the inset box. Both randomization and pseudonymization keep a dataset fairly intact while removing identifiable information. You can also retain the ability to re-identify such datasets by keeping a separate document in which you record the original and replacement values used for anonymization. This second document would, of course, require secure storage.

Hashing

One method for replacing identifiable data is to use hashing. Hashing algorithms take a value, reduce it to the bits that make it up, then perform a mathematical computation on those bits. For example, the algorithm SHA-1 takes the value "Marcy" and turns it into the 40-character hash "ad497d3b534378ac9070a0fc5aab73ad7252810c". A good hashing algorithm is a one-way transformation, meaning you can't tell anything about the input from the output. Good hashing algorithms also provide a nice output spread so that two different inputs are highly unlikely to result in the same hash value.

Hashing is a useful tool for pseudonymization as the algorithms are easy to use and identical inputs always yield the same hash value. Additionally, hashing can handle both characters (e.g. names) and numbers (e.g. ages) making it useful for all types of identifiable information. The downside to hashing, and pseudonymization in general, is that pseudonyms alone do not perfectly protect against re-identification. An example of this comes from the 2014 release of New York City taxi cab data where taxi license numbers where pseudonymized using the MD5 hashing algorithm. Unfortunately, license numbers have a fixed form and once someone identified that the MD5 hash was used, they were able to compute all of the possible license numbers and re-identify the data (Pandurangan 2014). In this case, it would have been better to assign a random number to each license before hashing. Note that hashing social security numbers also runs into this fixed-form problem, which is why hashing is not recommended for these values. However, when used properly and with other forms of anonymization, hashing is a useful tool for pseudonymizing information within a dataset.

No matter the technique, masking's effectiveness breaks down when dealing with indirect identifiers. For example, you lose useful information if you randomize age values in a dataset. Because these indirect identifiers often require anonymization, we cannot rely on masking alone. We therefore use de-identification as the preferred method of anonymizing data.

7.3.3 De-identifying data

Unlike blunt masking, de-identification aims to balance the risk of re-identification with preserving as much information in a dataset as possible. This usually results in a less distorted but still anonymized dataset. For example, de-identification might entail grouping ages into ranges instead of entirely removing this information from a dataset. This keeps the information usable but reduces the risk of identifying a particular person by age. These types of trade-offs are what make de-identification useful but also challenging to perform.

Techniques for de-identification

De-identification relies upon masking techniques but also adds two more techniques for anonymization: generalization and subsampling. Generalization involves making data points less specific, for example by changing birth date to birth year or replacing a specific address with the more general state or province. The goal is to be able to use the data but not include specific information that can identify someone. The other technique, subsampling, relies upon releasing a random subset of a dataset. The idea behind subsampling is that, even if you know someone is in the original dataset, it is more difficult to determine if a specific person's data is in the subset. Subsampling is a common technique for census data, as the dataset is large enough for a subset to be de-identified but still statistically relevant (El Emam 2014).

As with masking, you can use multiple de-identification techniques within one dataset. The choice of technique depends on the variable and the corresponding risk for re-identification. Therefore it is not enough to know techniques for de-identification, we must also examine methods.

Methods for de-identification

While there are many methods for de-identification, we will examine the two most common: lists and statistical de-identification (also called "risk-based de-identification" or "expert determination"). Lists are the more straightforward method, as they provide an explicit list of identifiers to remove from a dataset. This makes for easier, though not foolproof, anonymization.

One of the more well-known lists is the US's "Safe Harbor" provision in the healthcare data law HIPAA. Safe Harbor requires the removal of the following information from medical data:

- Names
- All geographic subdivisions smaller than a state, including street address, city, county, precinct, ZIP code, and their equivalent geocodes, except for the initial three digits of the ZIP code if, according to the current publicly available data from the Bureau of the Census:
    - The geographic unit formed by combining all ZIP codes with the same three initial digits contains more than 20,000 people; and
    - The initial three digits of a ZIP code for all such geographic units containing 20,000 or fewer people is changed to 000
- All elements of dates (except year) for dates that are directly related to an individual, including birth date, admission date, discharge date, death date, and all ages over 89 and all elements of dates (including year) indicative of such age, except that such ages and elements may be aggregated into a single category of age 90 or older
- Telephone numbers
- Fax numbers
- Email addresses
- Social security numbers

- Medical record numbers
- Health plan beneficiary numbers
- Account numbers
- Certificate/license numbers
- Vehicle identifiers and serial numbers, including license plate numbers
- Device identifiers and serial numbers
- Web Universal Resource Locators (URLs)
- Internet Protocol (IP) addresses
- Biometric identifiers, including finger and voice prints
- Full-face photographs and any comparable images
- Any other unique identifying number, characteristic, or code (US Department of Health & Human Services 2014)

It is important to note that applying the Safe Harbor list alone is not enough. The rule also stipulates that the "covered entity does not have actual knowledge that the information could be used alone or in combination with other information to identify an individual who is a subject of the information" (US Department of Health & Human Services 2014). Namely, if you know that a person in the dataset can still be identified after applying the list, you must modify the data until that person is no longer identifiable.

Safe Harbor and other such lists are popular because they are easy to use and a defendable standard for anonymization. However, such lists are not foolproof, as evidenced by the additional caveat to the Safe Harbor rule. Not every dataset will be able to use a list-based method for anonymization, as this depends on applicable laws, but it is a good starting point to approaching anonymization.

The other method for de-identification is the statistical method. This involves calculating probability of re-identification for a particular dataset and ensuring, via de-identification, that the probability is below a certain threshold. Probabilities differ for privately shared data and publicly shared data. The statistical method requires expertise in de-identification, making the exact details of this method outside the scope of what this book can cover. Still, we can examine the general strategies employed in the statistical method as a starting point for understanding how it works.

The goal of the statistical method is to remove direct identifiers and generalize indirect identifiers, balancing the risk of re-identification against minimal distortion of the data. The general strategy for de-identification includes (Corti 2014):

- Remove direct identifiers from the data
- Reduce the precision of a subject's age and birthplace
- Generalize text-based indirect identifiers, such as profession
- Restrict the range of continuous variables to eliminate outliers, such as for age
- Generalize location data
- Anonymize relational information within the dataset and with outside datasets

The final precision of variables and the amount of anonymization will ultimately depend on the dataset being de-identified, which is where expertise is required. If you are interested in this method of anonymization, there are lots of resources available to help you get started (El Emam 2014; ICO 2014; US Department of Health & Human Services 2014).

7.3.4 Other anonymization considerations

When anonymizing a dataset, there are a few other important considerations. The first is that you should document your anonymization process. By documenting the steps you take to anonymize your data, it makes the process reproducible and your data more understandable to others. Because anonymization inherently involves altering a dataset, one must have a basic understanding of the anonymization process in order to properly analyze the anonymized data. Documenting the anonymization process also gives you a clear record of the process, should concerns arise about re-identification of the data.

The second important thing to note is that anonymized datasets do not exist independently. Security expert Latanya Sweeney, mentioned earlier in this chapter, is also well known for proving this point. In 1997 she took a supposedly anonymized dataset of health records released by the state of Massachusetts and combined it with local voter rolls, which she purchased, to identify the governor of Massachusetts' health records. The point was to show the governor, who assured the public that their privacy was protected, that the dataset was not as anonymous as he thought (Anderson 2009). Unfortunately, using a second public dataset to re-identify an anonymized dataset is not uncommon. Therefore you must take care to properly anonymize data and particularly pay attention to indirect identifiers which can be combined with outside information to identify people within a dataset.

The story of the Massachusetts data highlights another point about anonymization and that is that the anonymization standards for publicly shared datasets should be stricter than for privately shared data. This is because publicly shared data is open to people trying to prove a point that your data is not properly anonymized. They do this by identifying the easiest to identify person in the dataset, which means at its weakest point your anonymization procedure must still be strong. Another benefit that privately shared data has is that you can dictate the terms of sharing, such as requiring the data recipient to store data in a particular way and to not attempt to re-identify the data. This does not eliminate the re-identification risk but does lower it, especially as compared to public sharing.

Finally, those dealing with public sharing mandates and sensitive data may wish to consult with an anonymization expert before releasing data. Alternatively, you can take advantage of exceptions in data sharing mandates for sensitive data and not share your data at all. Such exceptions exist because some types of data cannot be properly anonymized, but many can be prepared for public sharing with the proper care. As with security in general, this is an area where it pays to ask for expert help.

7.4 CHAPTER SUMMARY

Researchers deal with many types of sensitive data, the most obvious of which is data containing personally identifiable information. This type of data is usually subject to national law, governing how it can be collected, stored, and accessed. But even researchers who are not legally obligated to protect such data are ethically obligated to protect human subject data. Beyond these requirements for personal information, researchers may also wish to protect other types of data, such as intellectual property, and therefore use data security strategies.

When faced with sensitive data, there are lots of practices that you should use to keep your data safe. The first is following good computer security practices, including keeping your software patched and up to date, using a firewall and antivirus software, safely browsing the web, and using strong passwords. The next practice is to limit access by unauthorized individuals, including both physical and digital access. Many strategies for this exist and a particularly common one is to encrypt your files and devices. Finally, don't forget to properly destroy data and to make sure everyone knows the security procedures.

If you have data containing personally identifiable information, also consider anonymization. Anonymization transforms sensitive data into non-sensitive data, removing many legal obligations to handle data in a particular way. You have the option to perform masking, list-based de-identification, and statistical de-identification, among other methods, to remove direct and indirect identifiers from your data.

8

STORAGE AND BACKUPS

Billy Hinchen was a biologist working toward a PhD at the University of Cambridge until the day he returned home to find his laptop and external hard drives had been stolen (Figshare 2014a). Four years and 400GB of data were gone, with no way to recover them. Unfortunately for Hinchen, his hopes for a PhD disappeared along with his data.

I wish I could say that Hinchen's story is unique, but significant data loss is surprisingly common. There are many stories of graduate students losing theses on stolen laptops (Read 2010; Herald 2012; Wyllie 2012), students losing data on portal storage devices (Steinhart 2012), and even professors losing important research data (Katz 2011). It's just a sad fact that researchers lose important data every year to hardware failure, theft, computer virus, natural disaster, and accidental loss.

While it is not always possible to prevent loss of a storage device, it is possible to protect against ultimate loss of data by having proper storage and backup practices. This chapter covers those practices and makes recommendations to make your data safer from loss.

8.1 STORAGE

Everyone who has research data needs to be concerned about storage. Storage is a basic aspect of managing data but it's not always straightforward. There are a lot of options available and a lot of ways to go wrong. Still, with a few basic strategies you can make your data safer without having to be an expert in information technology. These strategies include using redundant storage, choosing good hardware, and making your comprehensive storage system fit your needs.

Note that all of these considerations are for short-term storage while you are actively working on a project. The considerations for short-term storage and for storage after you finish with the project and the data are very different. Long-term storage requires keeping files in a readable state and preventing corruption, which require different strategies than maintaining a working copy. Refer to Chapter 9 for strategies on long-term storage.

8.1.1 Storage best practices

The motto for storage and backups is that "lots of copies keep stuff safe" (LOCKSS). Keeping multiple copies means that even if one copy is lost, data is still secure in other locations. This concept is often put into practice by institutions that take part in formal "LOCKSS networks" (LOCKSS 2014), where they share storage with other member institutions to create redundancy. Such networks operate at a scale not realistic for the average researcher, often with five or more member sites, but the general idea of LOCKSS is one that can be leveraged for greater data security.

For most research data, a good rule of thumb is to follow the 3-2-1 backup rule (Leopando 2013; Levkina 2014). This guideline recommends maintaining three copies of your data on at least two different types of storage media with one offsite copy. Three copies is a good balance between having enough copies and having a manageable number of copies. The offsite copy is one of the most important copies as it protects against fire, natural disaster, and any other local event that might cause you to lose your data. This third copy is your just-in-case copy, while the two onsite copies are your main storage and backup. Finally, varying storage media compensates for the inherent risks of each storage type (see Section 8.1.2), such as a laptop's greater risk of being stolen.

Three copies is sufficient for most research data but you may want additional copies, for example if your data is difficult to reproduce or you are part of a significant collaboration or data sharing network. Here, extra redundancy reduces the risk of loss and makes it easier for collaborators to access the dataset. While a few researchers will need more copies, generally researchers should maintain three copies of their data, use two different types of media, and keep one copy offsite.

8.1.2 Storage hardware

There are many storage hardware options available to modern researchers and it can be difficult to know which are best for your research data. This section covers the many options and the pros and cons of each (Salo 2013; Digital Preservation Coalition 2014; The National Archives 2014; UK Data Archive 2014). Table 8.1 provides an overview of the most common hardware. Note that most of these options are not intended for long-term storage, as media can fail after five to ten years.

Table 8.1 Storage hardware options

Hardware	Rating	Notes
Personal computer	Good	Good when used with other storage
External hard drive	Good	Good when used with other storage
Local server/drive	Good	Good when used with other storage
Magnetic tape	Good	Good when used with other storage
CD/DVD	Acceptable	Cumbersome to use

Hardware	Rating	Notes
Cloud storage	Depends on product	Read the Terms of Service
USB flash drive	Do not use	Use only for file transfer
Obsolete media	Do not use	Remove data from old media as soon as possible

Good hardware options

The first storage option is your computer. This is the most obvious place to keep your research data as you probably already use your computer to store and manipulate data. Your computer is a decent primary storage location, but it does have some downsides. For example, it can be difficult to keep your data organized on your computer. Also, your computer may not be the most secure place to store your data, both in terms of storing sensitive data and protecting the data and hardware from loss. Laptops, in particular, are highly vulnerable to theft and there have been many unfortunate researchers who lost their only copy of their data when a laptop was stolen. Therefore, your computer must not be the only place you store data.

A second common option for data storage is the external hard drive. These have become especially popular in recent years due to increasing storage capacity and decreasing cost. The external hard drive is a good option for a backup system, especially when set up to back up your data automatically. The major downside of the external hard drive is that they degrade over time, particularly drives using flash media. Anticipate a lifetime of an external drive to be about five years (Pinheiro *et al.* 2007; Schroeder and Gibson 2007; Beach 2013). Therefore, it's a good idea to use an external drive in conjunction with a second offsite backup system.

Drives and servers supported by your local institution are another good place to store and back up your data. Not only are there often cost advantages, but using a local drive means that someone else is responsible for maintaining the storage system. The flipside of this is a question of competence – shared drives are useful if someone capable runs them, otherwise your data may be at risk. The other downsides of a local server are restrictions, as there are often access location and storage size restrictions on shared drives. Still, if you have access to storage on a local drive that is run by someone competent, this is a great option to use for one of the three copies of your data.

A final good option for a backup system is magnetic tape. Tape is currently not as common as the previous storage systems, but it is one of the better options in terms of reliability over the long term. Putting data on tape is a good way to maintain fidelity of the bits that make up your digital files. It is also possible to automatically back up to tape, making it easy to use. The one downside of magnetic tape storage is that it is slow to recover from. For this reason, it makes an excellent third, just-in-case copy of your data. It may be slow to get data back, but you at least can recover your data if you lose the other two copies.

The four hardware types just discussed – your computer, external hard drives, local drives and servers, and magnetic tape – make up the best options for the storage of research data, so long as you are maintaining three copies of your data. As each system has its own weaknesses, it's a good idea to use more than one type of hardware in your comprehensive storage strategy. Using several hardware options spreads your risk around, so you are less susceptible to the failures of any one type of hardware.

Acceptable hardware options

On the next tier down from the best storage options are CDs and DVDs. While optical disks do a good job of maintaining data (though they can become corrupt over time), they are a bit obnoxious to maintain as a dedicated backup system. This is because it is laborious to write to a CD/DVD and people are less likely to regularly back up data with a system that is awkward to use. Additionally, CDs and DVDs provide a snapshot of your files at one point in time, meaning you will generate many outdated copies of your data which can be difficult to organize. So, optical disks are not a bad option in theory but are less desirable for backup in practice.

Also in the secondary tier is cloud storage. While cloud storage can be safe and reliable, your data's security is highly dependent on who is providing the cloud storage. For example, cloud storage makes your data reliant on a company's business model. If the company folds or is bought by another corporation, there is no guarantee that you will maintain access to your data. More troubling are the terms of service for some cloud storage systems. Google Drive, for example, states the following in their terms of service:

> When you upload, submit, store, send or receive content to or through our Services, you give Google (and those we work with) a worldwide license to use, host, store, reproduce, modify, create derivative works (such as those resulting from translations, adaptations or other changes we make so that your content works better with our Services), communicate, publish, publicly perform, publicly display and distribute such content. The rights you grant in this license are for the limited purpose of operating, promoting, and improving our Services, and to develop new ones. This license continues even if you stop using our Services (for example, for a business listing you have added to Google Maps). Some Services may offer you ways to access and remove content that has been provided to that Service. Also, in some of our Services, there are terms or settings that narrow the scope of our use of the content submitted in those Services. Make sure you have the necessary rights to grant us this license for any content that you submit to our Services. (Google 2014)

Simply by uploading content to the Google servers, you give the company wide license over your data. While Google's terms of service are particularly encompassing, such liberal terms of service are not limited to Google alone. Therefore, you should always read the terms of service before putting your data in the cloud.

The other thing to look at is if the cloud storage service provides syncing or a true backup. Cloud storage that relies on a synced folder, such as Dropbox, only counts as one copy under the 3-2-1 backup rule, even if the data appears to live in both a local synced folder and the cloud. This is because if you delete a file in one location – either locally or in the cloud – this deletion propagates to other synced locations. So all instances of your synced folder can disappear if something accidentally happens to your data in one location. Unfortunately, such accidents have happened to both users of Dropbox (Čurn 2014) and Box (Tyan 2013). The alternative is to use a service like SpiderOak, which backs up versions of your documents over time. Even if you delete a file locally, a true backup will still retain old versions of that document in the cloud. Likewise, deletion in the cloud does not affect local copies of the data; the two copies exist independently. So opt for a true backup service over synced storage if you are using cloud storage for your offsite backup. However, if you want to access your data from multiple locations, synced cloud storage service is the better option.

There are good cloud storage options available if you are worried about your data, for example, services that encrypt data on your computer before transferring it to the cloud. That said, some data – such as data containing personal information – should never go into the cloud. In summation, I am not saying that you should not use cloud storage, rather that you should make an informed decision about your cloud storage system and always use cloud storage in conjunction with another storage option.

Not recommended hardware options

There is one more commonly used storage device and it occupies the tier below CDs, DVDs, and cloud storage – the USB flash drive. Flash drives are popular because they are inexpensive and can have a large storage capacity. These redeeming qualities do not counter the many downsides of USB flash drives, like the fact that small flash drives are easy to lose. Your data is not secure on something that can easily disappear. Additionally, flash media is one of the worst options for data fidelity; flash media becomes corrupt faster than most other systems. Finally, flash drives are an awkward backup system, making you more likely to forget to regularly back up files. I recommend only using USB flash drives for file transfer and using other systems for storage.

Finally, note that you should not be storing data on obsolete storage media. This includes everything from floppy disks to old computers gathering dust. If you currently have data on old media, copy the data onto a new device as soon as possible! This is because, as each year goes by, it gets progressively more difficult to find hardware to read old media. If you are in the position of having old media with no way to read it, you may be able to find help with your local IT support or library. The best option, however, is likely a business that specializes in data recovery. With more and more data being lost to obsolescence, data recovery is an area where we will see more options in the coming years.

8.1.3 Choosing storage

Armed with an understanding of the 3-2-1 backup rule and the landscape of storage hardware, it is now time to decide on the three options for your storage and backup system. Choosing the proper storage is a balance between hardware directly available to you, the systems that will make your data safest from loss, and other considerations about your data itself, such as privacy concerns.

If you have personally identifiable data or data you wish to keep secure above and beyond basic storage, refer to Chapter 7 for more information on secure data storage. Do note that there is a general trade-off between ease of access and security. The most secure data in the world is extremely difficult to access, while data that is easily accessed is not very secure. Keeping data secure means limiting access.

The first question to ask about basic data storage is how much do you need? There is a significant cost difference between maintaining 1GB of storage versus 1TB of storage, and having an estimate of the total size of your data ahead of time will ensure that you purchase the proper capacity. The goal is to have the right amount of storage for your needs but, when in doubt, err on the side of more storage in the event that you generate more data than expected.

The second question in choosing storage is what hardware options are readily available to you at your desired storage capacity? For example, you might not have the option of a local drive for storage but instead have a good cloud storage option through your workplace. Taking advantage of what is easily available is a great way to add up to three copies of your data. Remember that having some variety in your storage hardware can spread risk around to make your data safer.

Once you decide on the systems to use and their capacity, you must determine who will manage your storage. This can be you, an assistant, a local IT person, or an outside professional. There are many benefits to having storage run by a professional – including the technical expertise and a professional's greater focus on keeping storage systems running properly – and it can definitely be worth the extra cost to use professionally run storage. At the very least, you want someone to periodically check that things are running properly and to catch problems early, which protects data from loss.

8.1.4 Physical storage

While many modern datasets are digital, a significant amount of research data exists solely in the physical form. This category includes everything from analog measurements to physical samples to data recorded only in handwritten field notes. Just like digital data, analog data requires proper storage and maintenance. And just like digital data, we can ask some of the same questions to determine the proper storage system.

Firstly, you should consider the size of your analog data. Does it consist of two field notebooks or 10,000 bone samples? Physically, how large is the collection? How often will you be adding to the collection? Basically, you want to define the amount of physical space required for storage.

After determining size, you can decide on storage location. This can be anything from a drawer to a shelf to a filing cabinet to a whole room. For larger collections, you should also consider how you organize and label the objects. As with digital data, the best solutions are often those that are easily available to you, though you may need to also consider environmental considerations, such as excess heat and humidity and the need for security from locks and doors. Note that security through locks and doors is also useful for digital data (see Chapter 7 for more information). Good storage is not only the proper size, but it should be secure and not cause the deterioration of physical objects over time.

As with digital data, it is good practice to put someone in charge of managing the data. Most analog data can be managed by individual researchers, though large shared collections should have a central manager. This ensures that everything is properly cared for and stays organized.

Finally, don't forget about backing up your analog data. While objects such as animal specimens cannot be duplicated, it is often possible to make a copy of data in another form, for example by taking photographs of the physical item. Refer to the next section on backups for more information on backing up analog data.

8.2 BACKUPS

Imagine losing your research data. You go into work and find that everything is gone: your computer, your notes, and all of your work. Would you be able to recover? Could you recreate all of the lost work? Do you have the money to recreate all of the work? Do you have backups in place to restore from? How much would losing the main copy of your data set you back?

This might seem like an interesting thought experiment, but the total loss of research data is the unfortunate reality for numerous researchers, as demonstrated by the stories at the beginning of this chapter. Many of us know of people who have lost data or have read stories about someone losing their data. The horror stories are numerous but many of these events can be traced directly back to inadequate backup practices.

8.2.1 Backup best practices

If you get nothing else from this book, know that your data should be well backed up. Two backup copies is best, as per the 3-2-1 backup rule, but any backups are better than no backups. This is because there is absolutely no guarantee that your data is safe if you only have one copy. Keeping only one copy of your data is courting disaster. Therefore, you should have at least one backup copy.

For data storage provided by someone else – either a local IT person or a storage provider – do not assume that their backup systems are sufficient. The crash of the research-focused cloud storage system Dedoose in 2014 demonstrates that even professional systems using backups sometimes fail (Kolowich 2014). If the many researchers who lost data in the crash had kept local copies of their data, the data loss would not have been so devastating. So while you can take comfort in the fact that your storage provider backs up your data, you should still personally maintain a copy of your data just in case.

8.2.2 Backup considerations

The goal of a backup is to keep what is difficult to replace. For example, if it requires one week and almost no money to replace all of your data then your backup is less critical than if it requires three years and significant funding to replace your data. The more valuable the data, in terms of both time and money, the more important it is to frequently back up your data to a reliable system.

The first thing to know about backups – besides that you should ideally have two of them, one onsite and one offsite – is that you should do them automatically whenever possible. People are liable to forget to do manual backups or to put off complicated backup tasks, so having automatic backups makes your system more reliable. This in turn means safer data. Many systems, such as laptops, external hard drives and magnetic tape, can be set up to back up automatically. It is worth finding ways to automate your backup processes so that your data is better protected from loss.

The second thing to consider is how frequently to back up your data. The answer depends on the frequency with which you generate data and the value of your data; data that is generated quickly and is extremely difficult to replace should be backed up more frequently. Backup frequency also depends on how easy it is to perform regular backups using your backup system. Most researchers will want to back up their data on a daily, weekly, or monthly basis.

On top of frequency, you may need to decide on the type of backup to regularly perform: full, incremental, or differential. Full backups copy all of your files for every backup. Incremental backups only copy files that have been added or changed since the last backup, either incremental or full. Differential backups copy any files that have been added or changed since the last full backup. Depending on your backup system, you may have only one or two options of the three. If you have the choice, you should periodically perform full backups and use either incremental or differential backups on a more regular basis. This achieves a balance between resetting everything with a full backup and saving computational power with partial backups.

Finally, note that backup systems require the same considerations as for storage systems, in terms of choosing hardware, picking capacity, designating a manager, etc. Refer to the considerations in the storage section of this chapter for more advice.

8.2.3 Test your backups

One important aspect of keeping backups is that you must test them. This consists of two parts. First, periodically checking your backups ensures that they work properly and that your data remains safe even if you lose the main copy of your files. While you can scan through log files, if your backup system has them, it's still a good idea to check your actual backed up data in case there is an unseen error. For some researchers, checking backups a few times a year will be sufficient, but those with difficult-to-reproduce data or complex backup systems should check their backups more frequently. You should also check your backups after you change anything about your storage system to ensure that backups still function properly.

The second reason to test your backups is so that you know how to restore data from them. You don't want to be learning how to recover data from your backup when you've just lost the primary copy of your files. Learning to test restoring ahead of time will prevent a great deal of added stress should you lose your primary storage.

Like so many things in data management, a little forethought can make a huge difference. Nowhere is this more the case than for data backups. A few moments spent checking that your data is safe and retrievable can save you hours of lost work in the event of primary storage catastrophe.

8.2.4 Backing up analog data

Most people remember to back up their digital data but very few think to back up their analog data. This is unfortunate because analog data is also susceptible to loss, though often from different processes. Things like fire, flood, natural disaster, theft, and poor environmental conditions can all lead to analog data loss. And it's not just analog data that is susceptible. Many researchers forget to back up their written notes, which, if lost, can make the corresponding data practically useless.

The 3-2-1 backup rule is a good model to follow for analog data and notes, though it's not always practical. Digital data has the benefit of a simple copy and paste mechanism to reproduce bits whereas the duplication of analog data can be complicated. This means you should try to follow the 3-2-1 backup rule wherever possible and, when you can't, carefully maintain any data that cannot be duplicated. Additionally, if you have the option of only one backup copy, an offsite backup is preferable to two onsite copies.

It is easiest to back up analog research notes and other small flat objects, such as paper readouts, because they can be copied or scanned. You should therefore include plans to back up your paper-based documents in your overall backup strategy. Research notebooks should be scanned or photocopied once completely filled, while individual items can be copied on an as-needed basis. Many researchers will prefer digital scans to photocopies – again because it's easier to reproduce digital bits – but your backup strategy will depend on the reproduction tools you have readily available.

Three-dimensional objects are more difficult to back up, though you can capture information about such objects as images, inventories, descriptions, etc. Therefore, you should "back up" your three-dimension items with any information that could be used to continue your research or rebuild the collection if the originals are lost. For example, you could maintain photographs of bone samples as backups, used in place of the originals to assess shape, etc., or back up the inventory of plant specimens, useful for rebuilding the collection. With a little creativity in backups and a little care in managing the actual items, you can protect yourself from devastating loss.

8.3 CASE STUDIES

To finish this chapter, let us look at some case studies of storage solutions. The first case study involves a researcher, Sandra, who studies migration patterns of birds. Sandra has approximately 10GB of data on her personal computer and several handwritten research notebooks, all of which need backing up.

First, Sandra takes an estimate of the amount of data she has and the hardware options available to her. She finds that she has the option to use a local drive, run by her institution, for an onsite copy of her data and decides that cloud storage is a good option for her offsite copy. After reading a little bit about different cloud storage providers, she settles on SpiderOak, which offers encrypted backups in the cloud. SpiderOak automatically backs up her files in real time and she creates an automated weekly backup onto the local drive so that both of her backups are automated. Once everything is set up, Sandra tests each backup to make sure that it is working and to learn how to recover her information from the system.

Sandra also finds time to digitally scan her old research notebooks, putting the scanned files in with her other data on the two backup systems. She plans to back up future research notebooks in the same way. Taking stock of her complete system, Sandra is happy to find that she has three copies of her data and notes, two onsite and one offsite. She will manage all of her data going forward by periodically checking that all of her storage systems run properly.

In the second case study, we have a research group that studies plasma physics. In total, the group has eight members and almost 1TB of data and digital notes, most of which they share with their collaborators in Europe. The principal investigator of the group, Dr. Grange, oversees the entire group's data and facilitates the collaboration. She keeps the master copy of the group's data on a 2TB external hard drive, with files copied there from all members' individual computers.

This group's storage needs are a little different than in the previous case, as the data must be available outside of the group's home institution. Dr. Grange therefore works with an IT person to set up a server to provide FTP access to the data. After further consultation, Dr. Grange also decides to start weekly backups to magnetic tape at an offsite location. While she is ultimately responsible for the data, Dr. Grange uses some of her research funding to pay for IT support to keep the storage and backup systems running properly. Through a little planning and IT support, the physics group now has adequate storage and backup systems that are managed by a competent professional.

8.4 CHAPTER SUMMARY

Storage is central to managing data well, and the best practice is to maintain three copies of your data – two onsite and one offsite – to protect it from accidental loss. There are many hardware options to reach the target three copies, though the best options include your computer, an external hard drive, a local drive or server, and magnetic tape backup. If you have data on outdated media, move the information onto a new storage device as soon as you can.

Using good backup systems is just as important as using good storage. Choose systems that are easy to use and automatable. Test your backups to ensure that they work properly and to know how to restore from backup, should the need arise. Finally, it can be worth the cost of paying for professional services, as they are usually more reliable than storage and backups run by you.

9

LONG-TERM STORAGE AND PRESERVATION

On a night not too long ago, a research group was hanging out at their local bar, chatting about their work under the full moon. Something about the combination of the moon and the beer led them to think about a sleep study they conducted ten years previously and wonder if that data could tell them anything about the effect of lunar cycles on sleep. The researchers went back to their old data and, happily, found enough data points to correlate sleep and the lunar phase (Zivkovic 2013). Taking the recycled data and a new analysis, the researchers published their results in the paper "Evidence that the lunar cycle influences human sleep" (Cajochen *et al.* 2013).

This story is a great example of data reuse, a common occurrence in science. Data is frequently repurposed for new study or reproduced as the starting point for a related study. Yet for every success story of data reuse, such as for the lunar phase study, there are plenty of other examples of unsuccessful reuse attempts. Much of the time, reuse is prevented because data was not prepared for archiving or use after the end of the project.

The new digital landscape of research means that you can no longer place a dataset on a shelf and expect the data to be there in ten years. There are many things that can go wrong during this time period, including media failure (ranging from bit flipping – corruption at the binary-level of the file – to total device failure), software and hardware becoming obsolete, theft of a storage device, natural disaster, and poor management and planning. Anyone who has ever looked at old files knows that, even if you can read the hardware, there is no guarantee that you can use the files. Therefore, if you want your data to be usable in ten years, you need to expend some effort in preserving it.

While foolproof data preservation requires a high level of expertise, there are several simple things that any researcher can do to improve the odds that data will be usable in the long term. This chapter discusses those simple practices, covering topics related to: what data to retain, how to prepare data for long-term storage, and options for third-party preservation.

9.1 WHAT TO RETAIN AND HOW LONG TO RETAIN IT

Data preservation deserves a whole chapter of this book because many datasets have value beyond their original project. A dataset may be reused immediately or years later but, in order to reuse that dataset, it needs to be in a form that is usable. This often becomes a problem as data ages, but it is a problem that can be managed with a little forethought, which means preparing data for long-term storage and preservation.

Data preservation covers what happens to data after it is no longer in active use, including everything from adequate storage and backup to careful curation with active auditing. No matter what you do with your data in the long term, the transition from the short term to the long term occurs as you wrap up a project. While preservation is usually easier if you make decisions throughout a project with reference to the ultimate fate of your data, at the very least you should do work at this transition point to prepare your files for the future. This is the last opportunity before you stop using data and has the added benefit that you can see the full range of project information and make decisions about what to do with everything.

Making decisions about your data before you put it into long-term storage is actually one of the most important parts of preserving data. So, before you can even worry about the mechanics of preserving data, we must first determine what to preserve and how long to preserve it. This is the focus of the next few sections.

9.1.1 Data retention policies

Data policy is the first thing to consider when preparing data for the long term. Though you may not be aware of it, there is a good chance your data is subject to a policy requiring it to be saved for a certain period of time after the project ends. Such a policy is likely to come from your workplace and/or research funder, especially if you are funded by a large public funding agency.

For publicly funded research, there are some general trends in data retention periods. For example, the minimum retention time for federally funded research in the United States is three years, as set by the White House Office of Management and Budget:

> Financial records, supporting documents, statistical records, and all other non-Federal entity records pertinent to a Federal award must be retained for a period of three years from the date of submission of the final expenditure report or, for Federal awards that are renewed quarterly or annually, from the date of the submission of the quarterly or annual financial report, respectively, as reported to the Federal awarding agency. (White House Office of Management and Budget 2013)

For researchers in the UK, expected retention times are longer. Not every research council states a required data retention period, but the ones that do often require retention for at least ten years (University of Cambridge 2010). For example, the EPSRC policy on data retention is as follows:

> Research organisations will ensure that EPSRC-funded research data is securely preserved for a minimum of 10 years from the date that any

> researcher "privileged access" period expires or, if others have accessed the data, from last date on which access to the data was requested by a third party. (Engineering and Physical Sciences Research Council 2014)

In Australia, the recommended retention period is five years, as set by the National Health and Medical Research Council, the Australian Research Council, and Universities Australia in the "Australian Code for the Responsible Conduct of Research":

> In general, the minimum recommended period for retention of research data is five years from the date of publication. However, in any particular case, the period for which data should be retained should be determined by the specific type of research. For example:
>
> - for short-term research projects that are for assessment purposes only, such as research projects completed by students, retaining research data for 12 months after the completion of the project may be sufficient
> - for most clinical trials, retaining research data for 15 years or more may be necessary
> - for areas such as gene therapy, research data must be retained permanently (e.g. patient records)
> - if the work has community or heritage value, research data should be kept permanently at this stage, preferably within a national collection. (National Health and Medical Research Council *et al.* 2007)

So the general trend for retention time is three, five, and ten years from the project end or publication for the United States, Australia, and the UK, respectively. However, you should still consult policies from your particular research funder to know the start point and exact duration for the purpose of compliance.

It's also important to recognize that some types of data may require longer retention periods, as exemplified in the previous quote from the Australian Code for the Responsible Conduct of Research. In the United States, the Office of Research Integrity notes that data involved in misconduct investigations and patents requires longer retention periods:

> There may also be special requirements depending on the issue involved. For instance, in the case of research misconduct involving NIH funding, records must be retained for six years after the final resolution date of the case. As noted previously, it is also important to retain research data pertinent to patented inventions for the life of the patent in case the patent is challenged or if lawsuits should arise. (Office of Research Integrity 2014)

As in the United States and Australia, longer retention periods can be found for particular types of data in the UK. The Medical Research Council, for example, requires longer retention periods for clinical research data:

The MRC's expectations for research data retention are:

- Research data and related material should be retained for a minimum of ten years after the study has been completed.
- For clinical research undertaken in MRC research units and institutes, the MRC expects research data relating to such studies to be retained for 20 years after the study has been completed to allow an appropriate follow-up period.
- Studies which propose retention periods beyond 20 years must include valid justification, for example, research data relating to longitudinal studies will often be retained indefinitely and archived and managed accordingly. (Medical Research Council 2014)

Therefore, if you have medical data, patent data, disputed data, or important data for longitudinal research, expect your retention times to be longer than for normal data.

A final point to note about funder and national policies is that, often, data sharing and data retention policies are inextricably linked. In the UK, for example, there is a greater focus on data sharing than data retention, which prompts the longer retention periods; the expectation is that data sharing will occur via a third-party repository which will assist with the long retention time. The link between data sharing and retention mostly matters for those who will care for the data in the long term. For longer retention times, the recommendation is to use a data repository, which retains your data while also making it publicly available. See Section 9.3 and Chapter 10 for more information on data repositories and data sharing. The other option is to use the strategies covered in this chapter to manage your own data, though this is not ideal for data with required sharing and long retention times.

Your workplace is the other common source for retention policies that apply to your data. For researchers working in industry, the business owns the data and therefore will determine the required retention period. Some companies make this explicit while others do not, but the main thing is that the company is ultimately responsible for the long-term maintenance of the data. For researchers working in academic settings, the expectations are much less clear. Some universities have explicit policy on data retention but most do not. Here is an example of a university policy from Harvard:

> Research Records should be retained, generally, for a period of no fewer than seven (7) years after the end of a research project or activity. For this purpose, a research project or activity should be regarded as having ended after (a) final reporting to the research sponsor, (b) final financial close-out of a sponsored research award, or (c) final publication of research results, or (d) cessation of academic or scientific activity on a specific research project, regardless of whether its results are published, whichever is later. (Harvard University 2011)

University policies often reflect national data retention policies, but sometimes require longer retention periods. If you are unsure if you have a retention policy at your university, check with your institutional research office.

The policies discussed in this section show that data retention policies are usually specific about how long to retain data. However, these policies are not specific as to what data to retain, applying broadly to any research data generated under the umbrella of the policy. Since retaining every piece of data is not a good use of resources, let us now look at a more balanced approach to data retention.

9.1.2 Common sense data retention

Policy will only take you so far with respect to data retention, so I recommend a common sense approach. Basically, err on the side of keeping more data for more time but don't keep everything. It is simply not reasonable to keep every piece of data in the long term. Besides, keeping every piece of data usable over many years would likely take resources that most researchers do not have. Therefore, you should be selective about what data to retain in the long term. Most data has a natural endpoint after which you no longer need to store it. For some datasets, such as important natural observations, this endpoint may not come in our lifetimes but a significant amount of data does not even need to be moved into long-term storage.

A good rule of thumb is to always retain data that supports publications or reports and any data that you may reuse. It is also recommended to keep data that is irreproducible, such as data tied to a time and a place. On the opposite side, you probably don't need data from failed experiments, data that you won't use again, and data that is easy to reproduce. If you don't know what to do with a particular dataset, err on the side of keeping the data. Still, most researchers will have at least some data that isn't needed after the end of a project.

Just as we can use common sense on what to keep, we can use common sense on how long to keep it. I recommend retaining data for at least five years, preferably ten or more. This is longer than some required retention times, but those retention times are simply not long enough. An example of why this is comes from a partial retraction from 2013 in which the authors could not provide data in response to suspected duplication in a figure (Oransky 2013c). In the absence of the original data, the figure was retracted. Two things are important to note about this retraction. The first is that it is possible that there was no mistake, but without the original data we cannot definitively know if the figure was correct or flawed. The second thing to note is that the paper was six years old at the time of the retraction. This implies that you need to keep your data on hand for at least six years, probably more, in case questions arise about your work. Thus, the recommended retention time of ten years or more is just in case you, or someone else, need the data.

The final thing to note is that the decision to preserve a dataset is not a final decision. If you decide to keep something that you realize is useless in four years, you are allowed to get rid of that data (just be sure that you're not in obvious violation of applicable data policies). It is actually a good idea to go through old data every few years and cull what is no longer needed. This saves resources and ensures that you are not hanging onto a bunch of useless datasets.

The culling process is one that you should repeat at the end of your target retention period. For example, if you plan to retain your data for ten years, at the end of the ten years you should go through your data to see if you still want to save anything. At this point, you have transitioned from "need to save" into "want to save", so the decision on what to keep is entirely up to you. You can throw it all away, keep it all, or do something in-between – do whatever you think is best.

Deleting files

Most digital files can be deleted from a computer or hard drive in the normal way, but sensitive data requires a more thorough process. This is because standard deletion only removes the reference to the data's location in the computer's memory and not the actual files. Thus, the computer cannot return the file if you ask for it – as the computer does not remember that the file is there – but you can still read the file by scanning the hard drive. Refer to Chapter 7 if you need to perform true deletion on your files.

9.2 PREPARING YOUR DATA FOR THE LONG TERM

Whenever I talk to researchers and graduate students about long-term storage, I always tell them my own story – that within three years of finishing my dissertation, I no longer had access to my research data. While I'm sure I could find copies of the files somewhere, that doesn't mean I could actually use the data. The issue is that I never prepared my data for life after my PhD lab, meaning that my files only worked in a software program I no longer had access to and I left the only copy of my laboratory notebook with my PhD adviser. If only I had taken a couple of hours before I graduated to process my data, I would not be cut off from these valuable files.

I'm lucky that I work in data management now, because it means that I don't really need that data. But many researchers don't have that option; it's either recover the data or spend precious resources recreating it. There is a third option, however, and that is preparing your data for the long term so it is easy to reuse when you need it.

Since we already examined what to save, we are now ready to discuss how to actually save data for the long term. You do this by keeping datasets in a readable format, ensuring that datasets remain interpretable, and performing proper management to prevent data loss. These are all things you want to think about when you wrap up a project and move your data into long-term storage.

9.2.1 Keeping files readable

One of the biggest challenges of keeping data in the long term is the rapid evolution of technology. Much of the software and hardware we used ten years ago is now obsolete and it's incredibly difficult to extract data of that age into a usable format. For example, if you have a Lotus 1-2-3 file on a floppy disk, it's going to be quite a challenge to read that data. The better thing to do is plan ahead and put your files in a format that is likely to be more readable in the future. This saves a mad scramble for when you need the files later.

The process of keeping files readable is actually an issue of both software and hardware. Both can become obsolete and both have their own issues in the long term. Let us look first at the software side of the problem.

File formats

One of the big challenges with keeping data for the long term has to do with the format of a digital file. There are an amazing number of file formats available, especially for different types of research data. This becomes a challenge in the long term because you often need specialized software to open files. Compounding the challenge of variety is the issue of version. As companies upgrade software packages, they often upgrade file formats to the point where sometimes the new version of the software won't open files from several versions ago. Finally, it's a reality of the software industry that products come and go, so you may end up with a file format from a software package that no longer exists. All these things pose challenges to opening older files.

To counteract these challenges, one must plan ahead by choosing good file formats in which to store data. Obviously, it's best if you can pick a good file format at the beginning of a project for data collection, though many researchers do not have that option. Still, it's worth converting important files at the end of a project as you prepare data for the long term.

The key to picking a good file format is to look for something open, standardized, well documented, and in wide use. Basically, you should avoid obscure and proprietary file types wherever possible. If only one or two pieces of software open the files and neither is free, consider using a different file type. Table 9.1 contains several examples of file formats and their preferred alternatives. The list of preferred file types includes formats that can be opened by multiple software programs and are likely to still be supported going forward. There aren't strict rules for picking the ideal file format, so use your best judgment on your available options. When it doubt, opt for formats that are in wide use, as lots of people will need to open these files in the future and there is a higher probability of support.

Table 9.1 Recommended file formats

Data type	Recommended formats	Description
Text	.txt	Plain text format
	.rtf	Rich text format
	.xml	eXtensible Mark-up Language
	.pdf	PDF

Data type	Recommended formats	Description
Tabular	.csv	Comma-separated values
	.tsv	Tab-separated values
Image	.tif	TIFF
	.svg	SVG
Audio	.mp3	MP3
	.wav	WAVE
Video	.mp4	MPEG-4
	.avi	AVI
Geospatial	.shp	ESRI Shapefile
Database	.dbf	DBF

References: National Digital Information Infrastructure and Preservation Program 2013; Corti 2014

While you should definitely convert less desirable file types into preferred file types, it's a good idea to retain data in both file formats over the long term. This is because you can lose information, like formatting, during the conversion process. Retaining data in both file formats lets you use the original format while you still have access to the software and have a fallback copy of your basic dataset when you need it.

Note that, just as with choosing what data to keep, choosing file formats is not a final decision. If you notice the general population moving away from a particular software package, or you decide to switch software packages in your lab, look to update your old data to the new file formats. For example, Lotus Notes and WordPerfect were replaced by Microsoft Word as the "standard" way to create text files and, for best support, your older text files should follow these major successions in software. It's much easier to convert data while things are in flux than after the older software becomes obsolete because you have better access to the necessary software at this point in time. So as you periodically review your old data, check to see if you need to update any file formats.

The final recommendation for dealing with unusual file formats over the long term is that sometimes you just have to maintain a copy of the software. This can happen because there is no open alternative, or converting to an alternative file format results in the loss of important information. Saving software is less ideal because it means you must retain a working copy of the software instead of just storing a file. I know of labs that maintain old computers with old operating systems just to keep one software program working, which is a headache. Saving software is more resource intensive than converting files, so I do not recommend it unless there are absolutely no options for converting to a better file type.

Hardware

Once you have your files saved in a good format, you should think about the hardware that stores the files. Just as file formats can become outdated, so can particular types of media (think floppy disks). Additionally, media has a tendency to become corrupt over time, taking important files with it. A case in point, the backup storage company BackBlaze examined their drives and estimated the average life expectancy of a hard drive to be six years (Beach 2013). So if you want to store data files for ten years, you must think about and periodically update your storage hardware.

A good rule of thumb is to update your storage hardware every three to five years. This is roughly the expected lifetime of your computer and a good guide for other media as well. When choosing new hardware, I recommend using the current popular technology to ensure you will have good hardware support in another three to five years. With new hardware in hand, simply move your files from the old hardware to the new; note that you can use checksums to monitor the transfer and to ensure that files have not become corrupt on the old media (see inset box). This is also a good time to check for outdated file formats and to cull datasets you no longer need.

The final point to note about hardware for long-term storage is that you need both storage and backups. I'm less fastidious about checking backups for long-term storage, but you still want to have them – particularly if you have important data that you plan to reuse. Just because your data has gone into long-term storage doesn't mean that backups stop being necessary.

Checksums

If you are concerned about file corruption from media degradation or bitrot (corruption caused by bit flipping), you will want to look into checksums. Checksums are a useful tool that allows you to tell if a file has changed. Note that checksums are computationally related to hashing, which was discussed in Chapter 7.

Checksums work by performing a calculation on the bits making up a file and outputting a value of a specific length. For example, the MD5 algorithm changes the value "42" into the checksum "a1d0c6e83f027327d-8461063f4ac58a6"; this algorithm takes a value of any size, from zero to the bits that make up a large file, and returns a 32-digit checksum. For a specific input, checksum algorithms always return the same output. This can be leveraged to tell if a file has changed, as even a single bit flipping will result in a different checksum value for the file.

To use checksums on your research data, you should run an important file through a checksum algorithm to compute its checksum value. Do this at a point when you will no longer edit the file, such as when the raw data is created or the analysis is complete. Store the checksum in a separate file from the data. At this point, you can periodically run the checksum algorithm against the file and check the result against the stored checksum value. Not only is this useful for monitoring files in long-term storage, but checksums can also be used to ensure no bits were lost during file transfer.

9.2.2 Keeping datasets interpretable

Having the ability to open a file five years after its creation is not enough to ensure that someone can actually use that file. This is because you need context in order to understand how the file came to be and how to use the file. Without this context, you can't be sure if the file is the one you are looking for or what particular parts of the dataset mean. Basically, you must have good documentation to reuse older datasets.

Improving your documentation

The need for good documentation is keenly felt in the years after a dataset is created. Even if you don't need good documentation in the weeks after generating the data, details get lost to memory over time. Additionally, it is not uncommon for a person who is not the dataset collector to be the one using the five-year-old data. From experience, I know that it is incredibly frustrating to be the person going through someone else's old data and trying to piece fragments together because the documentation is not sufficient. Therefore, an extremely important aspect of keeping data usable for the future is making sure that data is properly documented.

While the best time to create documentation is when you acquire a dataset, it's still possible to improve documentation at the end of the project when you are preparing data for the long term. In particular, this is a good time to look for holes in your documentation, as you have the ability to look at everything comprehensively. Plan to spend some time reviewing your documentation as you wrap up a project in order to make sure it is sufficient going forward. Refer to Chapter 4 for more guidance on improving documentation.

Backing up your documentation

The other concern for documentation in the long term is keeping copies of your documentation alongside your data. Documentation should be stored and backed up just as data should be. Most people remember to do this for digital notes, but often forget about backing up written notes. This is a shame, as data without documentation can be difficult or impossible to reuse.

Backing up handwritten notes is not particularly difficult, but requires a little time spent in front of a copier or scanner. Taking photographs of each notebook page is another good backup option as inexpensive digital cameras are increasingly available (Purrington 2011). Backup is best done upon the completion of a research notebook or the completion of a project. It certainly takes time, but a few extra minutes spent copying can mean the difference between having useful data and not being able to use your data in the future. Additionally, if you choose to scan your notes for a digital backup instead of a physical one, you can now store your notes directly with your data and easily make extra backup copies.

Backing up notes is particularly important for student researchers, who often leave their research notebook with their research adviser upon graduation (research notebooks usually belong to the university or principal investigator). Students who do not make copies of their notes prior to leaving the institution often lose the ability to use their data in the future, even if they keep the data files. I highly recommend that all students, with the permission of their adviser, spend some time at the copier before they graduate if they want to ensure continued access to their research notes and the ability to reuse their data.

9.2.3 Long-term data management

The requirement to manage your data in the long term doesn't suddenly stop once you prepare the data, though it does get a bit easier at this point. The key is to periodically check the data and update things as necessary. You should plan to check on your old data every year or two for the entirety of the data retention period.

To properly manage things in the long term, check your data and update file formats, hardware, backups, documentation, and anything else that is needed. Essentially, you should undertake the same data preparation steps already outlined but on a smaller scale and with the latest technology. For example, as certain file formats become less common, your data checks may include converting file types for certain datasets. Basically, you want to head off issues before they become serious problems, like having files on outdated or corrupt media.

As you check your old data every year or so, ask yourself the following questions:

- Do you still know where your old data is?
- Has the data become corrupt?
- Are your backups in good order?
- Do you need to update storage hardware?
- Are the file formats obsolete?
- Can you still understand the documentation?
- Are you maintaining older data that you can now discard?

The answers to these questions will determine the steps needed to put your data back in proper order. All of these checks are something you can do, but it is also possible to designate the role of long-term data manager to another person. Either way, someone should be in charge of checking on the data in the long term.

9.3 OUTSOURCING DATA PRESERVATION

The good news about keeping data usable in the long term is that there are a growing number of options for outsourcing this task to others. Increasingly, libraries, non-profit organizations, and scholarly organizations are creating the infrastructure necessary to preserve large amounts of data for the long term. Look for the terms "repository", "data curation", and "digital preservation" to discover if these services are available to you. Table 9.2 lists several repositories that offer data preservation services.

Table 9.2 A selection of recommended data repositories for preservation

Repository	URL	Description
Dryad	www.datadryad.org	Predominantly biology repository with curation and retention
ICPSR	www.icpsr.umich.edu	Social and political science repository with curation and retention
UK Data Archive	www.data-archive.ac.uk	Social science and humanities repository with curation and long-term retention
Figshare	www.figshare.com	General purpose repository with long-term retention
Local university repository	N/A	Check to see if a data repository is available at your institution

Be aware that data often requires some amount of preparation for third-party preservation. This comes back to the fact that it is not worth preserving information that is not in a usable form. Sometimes this preparation work means changing the file format, but often it involves creating documentation. The exact format for documentation depends on where the data will be preserved, as many repositories use a particular metadata schema to aid with indexing the data. Be prepared to at least minimally describe your data when you pass it off to a third party for preservation.

The other thing to note about repositories is that, in addition to preserving data, most repositories' main purpose is to make data publicly available. It does not make sense for a third party to expend resources on preserving your data unless there is the greater benefit of letting others see and use the data. This means that many preservation services are data repositories, providing access to research data in addition to preserving the data for the long term. If you are worried about sharing but still want assistance with preservation, many repositories offer embargoes. This means that data goes into the repository immediately but is not made publicly available for a fixed time, such as one year. This can be useful if you want your data preserved but are not ready to make it available to others yet. Chapter 10 provides more information on data sharing repositories.

So, given a particular repository, how can you tell if it will help you preserve your data? This is an important question because the preservation of original research data is still a fairly new type of service and not all repositories offer this service. Additionally, new repositories are being developed while older repositories sometimes become too much to maintain and it can be hard to tell the good from the bad. So here is a process I use for evaluating repositories for preservation. The first question to ask about a repository is who supports it. Do you recognize the group or is the repository associated with a national organization or research funder? A repository from a group with a strong research mission is likely to be well supported over time. The second question to ask is how long will the repository support the data? This is not always answered explicitly, but a good sign is if the repository relies on

"LOCKSS", "CLOCKSS", or "Portico" for preservation. These services ensure that data remains available even if the repository is no longer supported. Another thing to look for is how the repository processes datasets. For example, the repository Dryad does the following checks on datasets:

> Dryad performs basic checks on each submission (can the files be opened? are they free of viruses? are they free of copyright restrictions? do they appear to be free of sensitive data?). The completeness and correctness of the metadata (e.g. information about the associated publication, the date on which any embargo is to be lifted, indexing keywords) are checked and ... copies may be made in different file formats to facilitate preservation. (Dryad 2014a)

All of these things help ensure that the dataset is, and remains, in a usable format and are good checks to look for in other repositories. Finally, it never hurts to ask other researchers and research librarians what they recommend for preserving data in your field.

9.4 CHAPTER SUMMARY

Between required retention of your research data and the possibility that you may use it again, your data should be taken care of in the long term. While various data policies require retention periods of three to ten years, it is good practice to maintain older data for at least five years, preferably ten. Keep data that supports publications and reports, data that you may reuse, and data that is irreproducible. To optimize resources, not all data needs to be saved for the long term.

Once you decide on what data to save and how long to save it, you should put the data in a format that ensures longevity. This includes using open and commonly used file formats, staying up to date with hardware, and retaining good documentation. You should go through your data every few years to update things as necessary and to cull data that is no longer needed.

While you can maintain your own data in the long term, another option is outsourcing preservation to a third party. This is ideal for data with a required retention period and required data sharing, but is also useful if you don't want to dedicate resources to preservation. Many repositories that preserve data are now available, including numerous repositories at universities, so there are many options for getting assistance with the long-term retention of your data.

10

SHARING DATA

In 2010, Harvard economists Carmen Reinhart and Kenneth Rogoff published a paper linking economic growth with austerity practices during an economic downturn (Reinhart and Rogoff 2010). This research was then used to justify many austerity policies and practices during the concurrent global recession. Unfortunately for many country's economies, the analysis in the Reinhart and Rogoff paper was wrong, meaning austerity was not the correct response to the economic downturn.

The person who discovered the paper's calculation errors was Thomas Herndon, a graduate student at the University of Massachusetts–Amherst, who examined the paper as part of a class assignment. When he could not reproduce Reinhart and Rogoff's results from publicly available data, he contacted the authors to get access to the spreadsheet they used for their analysis. Upon receiving the spreadsheet, he discovered missing data and incorrect analysis in the spreadsheet which, when corrected, did not support the link between austerity and economic recovery (Alexander 2013; Marcus 2013e; Stodden 2013).

This story is a powerful example of why data sharing is necessary, as this paper had such a large – and unfortunately negative – impact on so many people. It also demonstrates that, while reproducibility is a pillar of scientific research, studies are often not reproducible from a published article alone. Reproducibility is one of the main reasons behind data sharing but definitely not the only one. This chapter examines why data sharing is becoming increasingly expected, the several varieties of data sharing, and the intellectual property issues that exist when making data available to other people.

10.1 DATA AND INTELLECTUAL PROPERTY

Before we get into the practical aspects of sharing data, it's helpful to examine the landscape in which sharing exists. The act of sharing content with others occurs within a framework of intellectual property, which includes the rights you have (or don't have) over any original content. In this case, the original content is data.

The types of intellectual property that are most applicable to data are copyright, licenses and contracts, and patents. This section provides an overview of these three types of intellectual property. You should also be aware that you may have local institutional policies that cover these areas. Many universities, for example, stipulate that they own any patents derived from research conducted at the university. Additionally, institutions sometimes claim ownership of any copyright on your data or even ownership of the research data itself. So you should both read through this section and consult your local policies to best understand how the framework of intellectual property affects your data sharing.

Finally, it's important note that I'm principally writing from the perspective of the United States systems of copyright and intellectual property. While I will mention the regulations of other countries, it is simply not possible to comprehensively cover global intellectual property laws in a small amount of space. As such laws vary subtly between each country, it is best to consult local legal experts if you have questions about intellectual property and your data.

10.1.1 Data and copyright

Copyright often seems unclear when applied to research data. This is because copyright, by definition, is something that covers fixed expressions of creative works. Essentially, you can only copyright an original creation that you record in an enduring format. So you cannot copyright a drawing you made in the sand but you can copyright the photograph of it. Based on this premise, copyright does not apply to facts as they are not fixed expressions of creative works. Facts, such as "the sky is blue" and "the United States is made up of 50 states", are simply not creative. This affects research data, as data is often comprised of facts and is thus not copyrightable.

When copyright does and does not apply to data

Where data is a collection of facts, recorded faithfully during data collection, it does not benefit from copyright in most countries. This applies to a range of things from natural observations to raw experimental results. Essentially, any set of data that someone with similar equipment could collect is likely not copyrightable (SURF 2014).

The fact that facts aren't copyrightable does not mean that some types of research data aren't fixed expressions of creative works, especially when we take a broad view of the definition of data (see Chapter 1). Datasets consisting of images and videos, for example, likely enjoy the benefit of copyright where a spreadsheet would not. The act of selecting a field of view to capture in an image may be enough to qualify for copyright even if you are only using the image as data.

Creative collections of facts can also benefit from copyright. "Creative" is the key quality required for copyright, meaning that you cannot copyright a non-original collection of facts (Levine 2013). This distinction comes from the US Supreme Court, which in 1991 ruled that a telephone book cannot have copyright because it is not an original compilation (Anon 1991). Even though creating a telephone book requires effort – what the Supreme Court called "sweat of the brow" – effort alone does not qualify for copyright in the absence of originality. So where research data is a collection of facts that have been curated, such as cleaned up and selectively compiled from multiple datasets, copyright can exist.

There is no magic formula to determine when data has copyright, especially as the details of copyright law vary from country to country. Generally, visual and textual data are more likely to be copyrightable than raw numerical data, but the best way to determine if copyright applies to your data is to ask a legal expert.

If your data falls under copyright

When copyright does apply to data, it generally comes into existence automatically. You do not need to apply for copyright in most countries. Simply by fixing a creative work into an enduring format, the creator immediately enjoys copyright protection on that data. This gives you the exclusive rights for a fixed amount of time to reproduce the work, make derivatives of the work, distribute the work, and publicly perform or display the work, among other possible rights. Copyright also gives you the right to share these rights with others or transfer them completely. The duration of the copyright depends on the country. For example, the copyright term in the US is currently your lifetime plus 70 years after death. The UK has a similar copyright duration for literary and artistic works – which most research outputs fall under – with different durations for sound recordings (50 years), typographic arrangements (25 years), and databases (15 years) (Corti 2014). When you hold copyright over data, you have the same rights as when you hold copyright over any other type of creative work.

Content for which copyright does not apply or has lapsed exists in the public domain. Being in the public domain means that anyone has the rights to do whatever they want with the content. Still, having the rights and having the ability to use public domain content are not the same thing. If you do not make your data available, others cannot use it even if your data exists in the public domain. It is only once you make such data available that others can exercise their public domain rights to use this content.

Database rights

One other important point about copyright and data concerns database rights, which are a set of rights akin to copyright that exists in some countries (like the UK and the European Union) and not in others (like the US and Australia). Database rights, where they exist, provide rights for a database independent of the copyright status of the database's contents. These rights only exist where significant intellectual work is put into "obtaining, verifying or presenting the content in an original manner" (Corti 2014). Database rights are worth knowing about if you have databases of research data and work in a country where such rights exist.

What this means for your research data

The simple fact is that data exists in a gray area of copyright and it's possible that your data does not benefit from such protection. That's not to say that you have no rights over your data, only that you cannot copyright your data. Your data may enjoy some of the other intellectual property rights covered in this section. If you think your data may be copyrightable, the best way to get a definite answer is to ask a legal expert. Even if your data is in the public domain, you still have a lot of control over your data; others cannot use data they do not have access to.

On the other side of this problem, copyright (or lack thereof) determines what happens when you want to reuse someone's original research. This is because where copyright is unclear, allowable reuse of the data is also unclear. Chapter 11 covers this issue in more detail.

10.1.2 Licenses and contracts

We can't discuss data copyright without examining data licenses and contracts. Licenses are standardized rights given by one party to another that can be broad or conditional depending on the license (Krier and Strasser 2014). For example, I can license a dataset that allows you to analyze my data on the condition that you not use the data for commercial purposes. Contracts often achieve the same purpose as licenses but take the form of tailored agreements between two parties. For example, we can agree to a contract wherein I give you access to my data to analyze and you promise to not share it with others. The fact that both parties promise to abide by the specific contract is an important distinction between a license and a contract. The other notable difference between licenses and contracts concerns how they are handled in court.

Researchers often encounter contracts when acquiring sensitive or proprietary data. There will be a contract between you and the data owner that covers your use, or sometimes your whole institution's use, of the data. Such contracts often include requirements for you to limit access to the data or handle it in a particular way. If the data owner fails to provide the data or the data receiver fails to keep the data private, the wronged party can sue for breach of contract.

You are more likely to find licenses on freely available datasets. These licenses stipulate that you can use the data if you agree to the license's conditions. In the case of licensed data, instead of mutually agreeing to terms, you often agree to the license by simply using the content. This is much like open source software's use of licenses, such as the GNU General Public License (GPL), to share and reuse code. Common licenses for shared data include those from Creative Commons (www.creativecommons.org) and Open Data Commons (www.opendatacommons.org). These two groups provide a number of licenses with different terms for researchers to adopt. Just as with contracts, there can be legal ramifications if you do not follow the terms of the license.

Section 10.4.6 describes data licenses in more detail but I do want to mention here that licenses are ideal for sharing data. Licenses are easy to use and can clear away potential confusion on whether the data is under copyright. Licenses are also standardized as compared to contracts, making it easier to merge and use multiple licensed datasets than datasets governed by a variety of unique contract provisions. Licenses streamline data reuse, particularly in the context of public data sharing.

10.1.3 Patents

While copyright and licensing directly cover research data, patents instead protect an invention supported by research data. The patent gives you exclusive rights to harness that invention for a set amount of time. You can also license or sell the patent to others during that time period. As with copyright, the duration of the patent can vary by country and the invention receiving the patent, but most patent terms are 20 years. In return for this period of protection, all patent applications are made public after filing and the intellectual property becomes publicly usable once the patent period ends.

Trade secrets

One other type of intellectual property worth noting is trade secrets. The prototypical example of a trade secret is the formula for Coca-Cola, though trade secrets can also cover procedures, inventions, and other valuable practices. Secrecy protects this type of intellectual property as opposed to having a legal monopoly during a finite time period, as with patents and copyright. So long as you do not disclose the trade secret, you have exclusive control over the information. Note that others are still free to independently discover your trade secret or reverse engineer it because trade secrets do not have the same legal protection as patents. Still, trade secrets can provide a competitive edge for a limitless period of time if the trade secret does in fact remain secret.

Disclosure is the key issue with trade secrets. There is no requirement to disclose your invention, as with filing a patent, to gain a monopoly. In fact, to be called a trade secret and protected from industrial espionage, you must put measures into place to protect the secret. This can include security systems, employee non-disclosure agreements, non-compete clauses, etc. Refer to Chapter 7 for more information on data security if you are dealing with trade secrets.

Just because data is not the object covered by this type of intellectual property does not mean that the patent's supporting data is not important to control in this case. In fact, it is usually more important to exercise control over data supporting a patent than data protected by copyright. The reason is that early disclosure of data can count as "prior publication" which nullifies a patent. So if you have data that will lead to a patent, do not share it before you submit your patent application. Some of the data security practices outlined in Chapter 7 can help with this.

The other data concern is that you should keep the data supporting the patent on hand for the life of the patent, usually 20 years from the filing date. The reason for keeping the data is that, in the event of a patent dispute, you want to have all of the data that supports the patent to prove the validity of that patent.

10.1.4 Intellectual property and data sharing

Copyright, license and contracts, and patents provide the framework in which data sharing occurs. When sharing data publicly, you need to worry about reuse rights under copyright and disclosure prior to patenting. You must also be aware of any licenses or contracts that apply to your data and think about licenses or contracts you want to use for your data. I recommend having a basic understanding of each type of intellectual property so as to know when they might affect your data sharing.

As data sharing occurs in several different forms – local sharing, research collaborations, and public sharing (each covered in subsequent sections) – the relevance of the different types of intellectual property can vary by situation. So let's examine how the three kinds of intellectual property impact the different varieties of data sharing.

Copyright matters most when publicly sharing data, insomuch as you need to be clear about what reuse rights you grant to others. Licensing dramatically helps in this situation. Copyright also applies to local sharing and collaborations, but rights are much easier to resolve because you can often directly converse with the person holding the copyright. The most important thing about copyright and data sharing is, if copyright applies to your data, you must be clear about what rights you give to others in any data sharing situation.

Licenses and contracts affect data sharing on two fronts. The first is that if you are using data under a license or contract, you must abide by the license or contract terms. So even if you wish to share the data, you cannot if data sharing is forbidden by the terms of the license or contract that applies to that data. The other concern is using a license or contract when sharing your original data. Consider using a contract if you wish to privately share sensitive data and need to place restrictions on data storage and use. Licenses, on the other hand, are strongly recommended for public data sharing to streamline reuse (see Section 10.4.6). Other, less formal arrangements can be made when sharing data with your co-workers and through collaborations.

In terms of patents and data sharing, the biggest concern is keeping data private prior to submitting the patent application. Sharing data, figures, research results, or even the idea of the invention can lead to the loss of the patent under prior-publication rules or through scooping by other researchers. So if you work in patentable research, do not share your data until the patent has been submitted.

10.2 LOCAL DATA SHARING AND REUSE

The principal type of data sharing is the sharing and reuse that happens locally and informally, either by reusing a co-worker's data or reusing your old data. The latter form of local reuse is sometimes referred to as "sharing with your future self". Science, by nature, is built on previous research and many laboratories use data from a previous study as a starting point or a baseline for the new study. So for many researchers, local data sharing is the first and most important type of sharing they will encounter.

Local data sharing is where you really find out if your data management practices are sufficient. Can a co-worker find your data? Can they understand what your data is? Can they use your data? If the answer to any of these questions is "no", consider making changes to your data management strategy. Local data sharing and personal data reuse are so common within research that they are some of the best reasons to manage your data well. They are also a great metric by which to judge if your data management is adequate.

The good news is that if you manage your data well, local sharing is fairly easy to handle. Many of the practices covered in earlier chapters make it easier for you to reuse your old data. In particular, using good documentation (Chapter 4), consistent file naming and organization (Chapter 5), having a record of the analysis process (Chapter 6), and maintaining data in a readable format and preventing corruption (Chapter 9) make it easier to reuse data well after you originally created and analyzed it. An added benefit in this case is that there is often a little personal or institutional knowledge about the data to cover any gaps in understanding old data. So even if you don't publicly share your data, consider data management from the perspective of reusing your data again in the future and you'll have goals for improving your practices.

10.3 COLLABORATIONS

The next step up from local data sharing is collaborations. Like local sharing, collaborations are also common within science but are not as informal. The benefit of the added step of consciously agreeing to work together means you have an opportunity to address how you will conduct the research and, importantly, how you will handle the data.

Collaboration involves slightly different and more conscious strategies than for local sharing, especially when you are actively sharing data between collaborators. In particular, data management planning plays an even greater role because deciding on data management strategies ahead of time can streamline data use (see Chapter 3). This is because when everyone uses the same systems, it becomes much easier to pass data back and forth.

One of the simplest things you can do to make collaborative data sharing easier is to have all collaborators use the same organization and naming conventions (see Chapter 5). Using the same conventions allows any of the collaborators to easily find data. Anytime you start a collaboration, work out logical naming and organization conventions at the outset and stick to them. Even if they aren't perfect conventions, having everything consistent from the beginning of the project pays huge rewards in terms of data findability later in the project.

The use of conventions extends to data collection. If multiple people in multiple locations will be collecting data for one larger dataset, you should plan out the format for that data before any information gets collected. Define the scope of variables, decide what the units will be, resolve any format issues, choose a layout for spreadsheets (see Chapter 6), etc. Strive for consistency so that you don't have to expend a lot of effort in data clean up later. Certainly, you can fix things if some people record length in meters and others in inches, for example, but it saves time and frustration if everything is consistent from the beginning.

The other important thing to discuss at the beginning of the collaboration is what will happen to the data at the end of the project. Who owns the data or will at least be the main manager of the data? Where will the data be stored at the end of the project? Who gets to make decisions about the data, such as further sharing, once the project ends? It's best to be upfront about such things so you don't encounter any surprises later that can sour relationships. Give your data the best shot at longevity by putting someone in charge of the data after the project ends.

Unfortunately, it's rare to find policy to guide you on data ownership in collaborations across different institutions. In lieu of policy, the best you can do is be upfront about how things will be handled. You can share ownership with key people making all the decisions or designate one person to handle everything. It doesn't matter what you decide so long as you have the conversation and work things out to the agreement of all parties. If you do not have this conversation, the assumption is that the project's principal investigator gets to make all the decisions about the data.

Beyond planning and deciding on ownership, everyone in the collaboration should follow good data management practices. As with local sharing, these include using good documentation (Chapter 4), consistent file naming and organization (Chapter 5), recording the analysis process (Chapter 6), picking the right storage options (Chapter 8), and maintaining data in a usable format after the end of the project (Chapter 9).

Finally, with collaborations, it's a good idea to designate a data manager who will oversee data storage and organization while the project is ongoing and can help prepare the data for long-term storage at the end of the project. The goal is to combat entropy by keeping things organized, preventing data from arbitrarily spreading across multiple computers, and ensuring that backups work. These are often the problems encountered when managing a single person's data, and which only get worse with a group's data. The larger the collaborative project, the more likely you'll need someone in this role, to the point where the managers of significant projects involving multiple sites should think about hiring an information professional.

10.4 PUBLIC DATA SHARING

This book cannot have a chapter on data sharing without devoting a significant section to public data sharing. Making data publicly available is a growing trend within scientific research and will soon become a normal part of the way we disseminate research results. Public data sharing is still not widely understood and adopted, but it will become important, so let us start by examining the reasons for this movement before getting into the details of how to share your data.

10.4.1 Reasons for public sharing

In 2011, the field of psychology was shocked to learn that researcher Diederik Stapel was under investigation for research misconduct (Verfaellie and McGwin 2011; Bhattacharjee 2013). Up to that point, Stapel had been a leader in the field with many high-profile papers on priming, the idea of unconsciously introducing concepts to see if it affects a person's behavior. For example, Stapel conjectured that a messy environment made one more likely to discriminate against others and that meat eaters were more antisocial than vegetarians (Stroebe and Hewstone 2013). Unfortunately, Stapel had no solid evidence for his priming studies; the inquiry committee at Tilburg University, where Stapel worked, found that he committed data fabrication in the majority of his publications. Following the committee's report, Stapel lost his job, his PhD, and had over 50 papers retracted (Oransky 2013f).

The depth of Stapel's data fabrication makes this story significant, but it is the reaction of psychology and other fields that makes it particularly notable. Since Stapel's data fabrication came to light, there has been an increased focus on reproducibility in all fields of science. This is backed up with funding, such as a $1.3 million grant for the Reproducibility Initiative (Iorns 2013; Pattinson 2013), and studies that reveal the lack of reproducibility in research – one investigation found that half of the cancer researchers studied have been unable to reproduce published results at some point in time (Mobley *et al.* 2013). Reproducibility is a cornerstone of scientific research and the current system is obviously not working.

The focus on reproducibility is central to the increased push for data sharing. This is because it is often impossible to reproduce results from a published article alone, as evidenced by the Reinhart and Rogoff example at the beginning of this chapter. Another way to say this is that "a scientific publication is not the scholarship itself, it is merely advertising of the scholarship" (Buckheit and Donoho 1995). In lieu of performing every study twice to ensure reproducibility, researchers and funders want to make studies more transparent. The thought is that by examining data, you can better tell if research is reproducible or not. While this won't catch every flawed study, closer examination of research data can uncover accidental errors, deliberate fabrication, and irreproducible results. This can only be a good thing for science overall.

Fueling this push for data sharing is the fact that it's easier than ever before to share data. The majority of research data is now born digital and can be easily distributed with the click of a mouse. So now you can ask a question, download a dataset, and perform your analysis in the space of an afternoon. Science, as a whole, is discussing data sharing so much because it's actually feasible to share data.

Both reproducibility and ease of sharing resonate with public research funders, who are tasked with promoting science by supporting the best research. The problem is that many funders are experiencing an increased demand for funding without an increase in their budgets. This makes data sharing particularly appealing to funders because sharing and reuse mean that more research can be conducted without spending more money. Adding to this issue is the increasing pressure to make the outputs of publicly funded research, both articles and data, freely available to the public that supported the work. All of these factors mean that many mandates for public sharing come from research funders (see Section 10.4.2).

Individual benefits of data sharing

Reproducibility, the prevalence of digital data, and demands on funders provided the catalysts for data sharing, but data sharing also offers benefits to the individual researcher. The first benefit is that published articles with shared datasets receive more citations than articles without shared data (Piwowar and Vision 2013). Piwowar and Vision found a 9% statistically significant increase in citations for biology papers with shared gene expression microarray data as compared to similar papers without shared data. This effect continued and increased for years after publication, with the almost ten-year-old papers in the study seeing a 30% increase in citations. So if you want more citations on your articles, share the data that supports that article.

You also get citations whenever someone uses your data. Data citations work just like article citations and you can put both the dataset and its citation count on your curriculum vitae. Data, like articles and code, is an important research product that you should claim to show research achievement. Section 10.5 goes into more detail on how to get credit for your shared datasets.

Another benefit of sharing is that it opens you up to new collaborations in ways not possible from articles alone. For example, a researcher using your published dataset for one project may need similar data for a second project, leading to collaboration. Data sharing and reuse takes you a step toward collaboration simply by mixing data from two different researchers. This can lead to more formal collaborations that benefit both parties.

Finally, as data sharing becomes more common, you have the opportunity to benefit from someone else's data. For example, if you begin a study based on someone else's research, you are able to directly use their data instead of starting your research from only a published description of the work. Using data from previous studies can save considerable time and effort in re-collecting the data. In other cases, it means you have access to data you would not otherwise be able to generate. This is a particular benefit for researchers at smaller institutions and in countries with less funding for scientific research. Anyone can directly work with information at the cutting edge of research when data is publicly available without cost. Researchers will see a lot of new possibilities that were not open to them before as data sharing becomes normal within research. For more information on finding and reusing data, see Chapter 11.

10.4.2 Sources of public sharing requirements

As already mentioned, funding agencies are the main source of sharing requirements. In the US, the National Institutes of Health (NIH) and the National Science Foundation (NSF) have been the major drivers for data sharing, but data sharing policies have extended to other federal funding agencies. In the spring of 2013, the

White House Office of Science and Technology Policy published a memorandum on public access (Holdren 2013) requiring all major US federal funding agencies to enact data management and sharing requirements for their grantees. This memo tipped the balance toward data sharing being the expectation when receiving federal funding in the United States. In the UK, Research Councils UK and the Wellcome Trust have been the major drivers of data sharing policy since the mid-2000s (Wellcome Trust 2010; Research Councils UK 2011). Their policies are generally stronger than those from US funders, calling for "researchers to maximise the availability of research data with as few restrictions as possible" (Wellcome Trust 2010). Taken as a whole, the general trend of federal research funders is to adopt data sharing requirements with many smaller funders following suit.

The other place you will encounter data sharing mandates is when you publish. Many journals now expect you to make the data that supports a published article available to others after publication and sometimes as part of the peer review process. Policy requirements vary by journal. Some journals, such as *Nature* (Nature Publishing Group 2006) and *Science* (Science/AAAS 2014), encourage making data available by request and depositing particular types of data into a repository, such as DNA sequences and microarray data. Other journals, such as the PLOS family of journals (Bloom 2013), will not publish your article unless you deposit your data, of any type, into a data repository. Currently, a limited number of journals have a data sharing requirement, for a list see (Strasser 2012; Dryad 2014c), but I expect more to adopt such policies in the future.

Freedom of Information Act

Providing research data under the Freedom of Information Act (FOIA), in countries where such legislation exists, deserves a special mention here. In both the US and the UK, FOIA requests can be used to acquire data from public universities. This type of data sharing is unlike anything else discussed in this chapter and applies to data that is ostensibly public record.

Freedom of Information Acts were created to allow citizens to request government information that is not otherwise publicly available. This is the main use for FOIA requests but requests have also been used to gain access to research data funded through government agencies. Most famously, dissenters used FOIA requests to try to get research data surrounding the email hacking of climate scientists, commonly referred to as "ClimateGate" (Corbyn 2010). Both the hacked emails and the University of East Anglia's response to the FOIA request were critically examined during the formal investigation after the incident. So if you work at a university and have federal funding, you need to be aware of FOIA and know how to respond to an FOIA request – particularly if you do politically controversial research.

The good news is that there are limitations to FOIA that prevent people from requesting all of your research records. First, requested records must be wholly or partially funded by the government. In the US, FOIA requests for data are limited to data as defined by the OMB Circular A-81:

> Research data means the recorded factual material commonly accepted in the scientific community as necessary to validate research findings, but not any of the following: preliminary analyses, drafts of scientific papers, plans for future research, peer reviews, or communications with colleagues. This "recorded" material excludes physical objects (e.g., laboratory samples). Research data also do not include:
>
> (i) Trade secrets, commercial information, materials necessary to be held confidential by a researcher until they are published, or similar information which is protected under law; and
>
> (ii) Personnel and medical information and similar information the disclosure of which would constitute a clearly unwarranted invasion of personal privacy, such as information that could be used to identify a particular person in a research study. (White House Office of Management and Budget 2013)

Requests for research data must also be specific to what data is being requested, meaning that you do not have to fill requests for data that does not exist. Additionally, if it is difficult to find or compile the requested data, you are allowed to charge a fee to the requester for time spent fulfilling the FOIA request. Other limitations to FOIA exist, with specifics depending on the requirements of your country's Freedom of Information laws.

If you get an FOIA request for your research data, the first thing you should do is ask for assistance. Do not handle this on your own. Contact your institution's legal or research support, who will help you respond to a request in the appropriate way. There are certain nuances of FOIA and you will want expert help in dealing with any requests for your data.

10.4.3 When and what to share

The nominal expectation for data sharing is that you will share the data that underlies a publication and make it publicly available around the time of publication. This means you don't have to share results before you are ready but it also means that others can review your work once it is published. Be aware that a few funders expect data sharing prior to publication. For example, UK funders "ESRC and AHRC (only in the case of archaeology) expect an offer of data to their data centres within three months of the end of the award. NERC expects data to be deposited as soon after the end of data collection as is possible, which may be well before the end of the award" (Digital Curation Centre 2014b). Additionally, while most journal data policies require sharing at the time of publication, a few want to see data for peer review. As always, refer to the specific policies that apply to you.

In terms of what to share, plan to share any data that supports the publication, unless applicable policies say otherwise. This means everything from data used in tables, data turned into figures, images that you performed analysis upon, etc. If someone will need a piece of data to reproduce your results, plan to share it.

The actual format of your shared data is up to your best judgment. For example, some scientists share raw data while others prefer sharing cleaned up and partially analyzed data. There are not yet best practices in this area, so share your data in a format that you believe will be most useful to others.

Open notebook science

Most researchers who share data make it available after publication but there is a small group of scientists who share data in real time. This practice, called "open notebook science", means that researchers immediately post data, notes, and analysis to the internet in the course of doing research. The idea is that having more transparency in the research process and engaging others before peer review makes for better scientific studies. The downside is that, by making data available before publication, these researchers can have their ideas scooped and published by someone else. Open notebook science also nullifies patents because open data sharing counts as prior publication. However, to the practitioners of open notebook science, the benefits of engagement and transparency outweigh the risks of losing ideas to others.

There are a few types of data that are common exceptions to sharing requirements. For example, many sharing policies provide an exemption for sensitive data, such as that containing personally identifiable information. The NIH includes such an exception in its sharing policy:

> NIH recognizes that data sharing may be complicated or limited, in some cases, by institutional policies, local IRB rules, as well as local, state and Federal laws and regulations, including the Privacy Rule ... The rights and privacy of people who participate in NIH-sponsored research must be protected at all times. Thus, data intended for broader use should be free of identifiers that would permit linkages to individual research participants and variables that could lead to deductive disclosure of the identity of individual subjects. When data sharing is limited, applicants should explain such limitations in their data sharing plans. (National Institutes of Health 2003)

Where anonymization of sensitive data (see Chapter 7) is not possible, your applicable policy will state if you are allowed to not share or if you must put data in a restricted repository.

Data sharing policy exceptions also exist where data supports intellectual property claims like patents. Research Councils UK, as part of its Common Principles on Data Policy, states that data "should be made openly available with as few restrictions as possible in a timely and responsible manner that does not harm intellectual property" (Research Councils UK 2011). Check your applicable policies for intellectual property exceptions that apply to your data.

In addition to data, the progress and reproducibility of research improves when researchers share code. This is especially true for studies that are based solely on computation, as the code represents the research methods. It can be extremely challenging to reproduce research based on personal code if you do not have access to that code. For this reason, there is a small but growing movement in support of shared research code (Stodden 2010; LeVeque *et al.* 2012). See Chapter 6 for more information on sharing code.

10.4.4 Preparing your data for sharing

The simple fact is that data usually requires preparation for sharing. This is because data that is in a form that is usable for you is not likely to be understandable to others. If you have ever picked up and tried to use someone else's data, you know that it can be difficult to understand, let alone use, data you did not create. So you must spend a little time preparing your data if you are going to share it.

One of the biggest preparation steps for data is quality control. Your data should be clean, consistent, and without errors. If you have tabular data, plan to follow the spreadsheet best practices laid out in Chapter 6. If you have textual data, be sure to use consistent labels and spelling. Also make sure that your units, data formats for each variable, and null values are consistent. You may not need to do a lot of clean up prior to sharing, but you should plan to double-check your data for quality at the very least.

Besides cleaning data, plan to set aside time to properly document any shared data. Data without documentation can be impossible to understand and use, so documentation is just as important as quality control for shared data. You can document within a dataset using comments and well-named variables, but data often benefits from external documentation. Consider using a data dictionary for datasets with many variables, a codebook for coded data, or a README.txt file for general documentation; Chapter 4 covers each of these types of document in more detail. You may also need to use a particular metadata schema or metadata fields, depending on where you share the data. For example, the repository GenBank lays out what metadata to include with the deposit of different types of genomic data, such as name, country, host, collection date, etc. for viral sequence submissions (NCBI 2011). Repositories requiring specific metadata usually provide support for putting your documentation into the correct format.

Once you prepare the data, you will need to make it available to others. If you are hosting your data on a personal website or uploading it to a repository, this will add a few extra minutes to the data preparation process. Plan to add some extra information, such as the main contact person for questions about the data and corresponding publications, during this step.

All of this preparation obviously takes time. One ecology researcher, Emilio Bruna, estimated that it took him ten hours to prepare and deposit his datasets with another 25 hours spent cleaning his code and making it available (Bruna 2014). He did expect the process to be faster and easier the next time he shares data. Still, data preparation is a time investment. However, I firmly believe that managing your data well from the start of a project makes it easier to prepare your data for sharing at the end of a project.

10.4.5 How to share

The best way to share digital data is to put your data in a repository. Many repositories are available for research data, with more being created every year. Putting your data into a repository has many benefits, the first of which is that once you submit your data you no longer need to worry about upkeep or maintaining access for others. The repository will ensure your data remains available. In addition to being hands off, repositories also provide a higher profile home for your data than a personal website. This is because, as repositories collect more and more data, people are starting to expect data will be in a repository. So putting your data in a repository means that your data is discoverable and thus is more likely to be cited than if your data is hosted elsewhere.

In conjunction with being discoverable, many repositories provide a citation for the dataset, a citation to the corresponding article, and have built-in tools for tracking metrics such as downloads and views. As attribution is just as important for data as it is for articles, these features reinforce good citation practices for data. See Section 10.5 for more information on tracking data citations and Chapter 11 for more information on citing data.

The other major benefit of data repositories is that they actively work to keep data available for a long time. For example, most repositories offer more permanence than website-hosted data through the use of Digital Object Identifiers (DOIs). DOIs are akin to web address URLs but always point to a particular object, even if that object's actual URL changes over time. Supporting the permanence of DOIs are features like robust backup networks and data curation. For example, the Dryad repository (Dryad 2014a) frequently curates data by converting datasets into more open file formats to maintain backup copies that are more likely to be readable in the long term. Chapter 9 goes into more detail on repositories that preserve data in the long term.

Data papers

Depositing data into a data repository is the typical way to share data but there is another noteworthy data sharing paradigm: publishing a "data paper". Data papers take sharing a step beyond depositing data in a repository by performing peer review on both the data and its documentation. The data paper consists of information about the dataset and gets published in a journal, while the data itself is either published alongside the paper or made available in a third-party repository.

Data papers look like traditional articles but differ in that a data paper describes only the dataset and not any scientific analysis or conclusions drawn from the dataset. Data papers do, however, receive citation in the same manner as a traditional article, which is why some scientists prefer them.

The purpose of the data paper is to provide greater documentation to an important dataset and add authority to the dataset through peer review – traits which make a dataset more likely to be reused. Therefore, consider publishing research data as a data paper when your data has high value to others. Many regular journals, such as *PLOS ONE* and *Ecology*, accept data papers as one of the many types of articles they publish while other journals, such as *Scientific Data* and *Open Health Data*, exclusively publish data papers. For a full list of journals accepting data papers, see these references: Callaghan 2013; Akers 2014.

Repository alternatives

The alternatives to using a repository are to make your digital data available by request or to host the data on your personal website. Neither of these options is ideal, as they require more work on your behalf. Not only will you have to respond to requests for data and keep tabs on where the data is, but you do not get the benefits of stability and easy citability that come with a repository. That's not to say that these aren't valid methods for data sharing, but when given an option using a repository is preferable.

The one exception to the recommendation against sharing-by-request is when you share sensitive data with restrictions. Sometimes it is not possible to completely anonymize a dataset and the only way to share the data is to place restrictions on their handling and reuse. For example, you might share on the conditions that the data not be shared further, human subjects in the dataset not be re-identified, and that the data must be stored in a secure environment. In this case, making your data available by request and making a specific agreement with the data reuser gives you the most control over sharing while keeping the data secure.

Sharing physical samples

Do note that some data sharing policies require physical specimens to be made available in addition to digital data. For example, the NSF Earth Sciences Division data sharing policy states:

> Investigators are expected to share with other researchers, at no more than incremental cost and within a reasonable time, the primary data, samples, physical collections and other supporting materials created or gathered in the course of work under NSF grants. (NSF Division of Earth Sciences 2010)

In this case, you can make physical samples available by request, as it is difficult to make physical objects universally available. There are some exceptions to this. The journal *Nature* particularly states that for "biological materials such as mutant strains and cell lines, the *Nature* journals require authors to use established public repositories when one exists" (Nature Publishing Group 2006). It's therefore best to consult any applicable sharing policies to know how to share physical samples. Depending on the type of samples you have and the policies that apply to your work, you have the option to place your materials in a public repository, send samples to anyone who requests them, or allow people to come and examine your samples.

Choosing a repository for your digital data

If you are using a repository for data sharing, you will have to decide which repository you will use. The good news is that there are many repositories available for a wide range of disciplines – see Table 10.1 for a non-comprehensive list of recommended repositories – but this can make it difficult to choose the best place for your data. So let us look at some strategies for choosing a good data repository.

Table 10.1 Select list of data repositories

Repository	Data types	URL
figshare	All	http://figshare.com/
Dataverse	All	http://thedata.harvard.edu/dvn/
ZENODO	Science	http://zenodo.org/
Dryad	Science, mostly biology	http://datadryad.org/
GenBank	Genome	http://www.ncbi.nlm.nih.gov/genbank
DataONE	Earth observational data	https://www.dataone.org/
GitHub	Code	https://github.com/
ICPSR	Social Science	https://www.icpsr.umich.edu/icpsrweb/landing.jsp
UK Data Archive	Social Science and Humanities	http://www.data-archive.ac.uk/
Local university repository	All	See local resources

If you want to share data that goes along with a published article, the first place to look for a repository recommendation is the journal where you published the article. A growing number of journals require data sharing and specifically recommend where you should put your data to facilitate the process. For example, the journal *Scientific Data* has a great list of recommended repositories to use when you publish in that journal (Scientific Data 2014). Some journals even facilitate putting your data in the repository, as with several journals that integrate with the Dryad repository (Dryad 2014b). Obviously not every journal will have repository recommendations, but it's a useful place to start.

The second thing to consider is where your peers share their data. Choosing a repository popular with your peers will make your data more likely to be discovered by others and more likely to be cited. This is just like publishing in a journal that's popular in your field – you put your content there because it will have a greater impact.

It is also possible that you have local options for data sharing in the form of an institutional data repository. This type of data repository is currently less common and is usually found only at large, research-intensive universities. For example, both Purdue University (Purdue University 2013) and the University of Minnesota (University of Minnesota 2014) have data repositories for researchers at those institutions. Be aware that some universities take part in consortial data repositories, such as the 3TU.Datacentrum (3TU.Datacentrum 2013) supported by Eindhoven University of Technology, Delft University of Technology, and University of Twente. If such a repository is available to you, it's definitely worth considering as a place to host your data, if only to take advantage of the local assistance.

In the absence of a local data repository, you can still ask your local librarian or data management support person for a repository recommendation. As data management and sharing becomes more prominent, many institutions offer support in these two areas. Since part of data management compliance includes knowing where to put data after the end of a project, many data management support people can help you navigate the field of repositories. Additionally, as librarians are experts in finding content including data, they can also offer help in finding a place to put your content.

Permanent identifiers

In choosing a repository for your research data, look for one that assigns a permanent identifier to your data. The most common type of permanent identifier is a DOI (digital object identifier), but you can also use a PURL (permanent URL) or ARK (archival resource key). The point of using a permanent identifier is that, while URLs often change over time, permanent identifiers should always point to the same object even if that object's URL changes. Thus, DOIs and their kin offer stability in the ever-changing landscape of the internet, making your data more findable and citable.

In the absence of any specific recommendations, refer to the list of repositories at re3data (www.re3data.org). This list breaks down repositories by subject and provides a short description of each repository. If you see a repository that might work, you'll want to do a little bit of vetting. First, look into who supports the repository. Do you know and trust this group? Second, does the repository do anything to aid data longevity, such as provide DOIs, participate in a backup network, or perform data curation? See Chapter 9 for more information on vetting a repository for preservation. Finally, you can also look at whether the repository provides support for easy citation, tracking metrics, and anything else that makes the repository easier to use.

There are many good repositories out there and there are many resources available to help find the best one for your data.

10.4.6 Licensing shared data

One thing you might encounter when publicly sharing your data is a license. How this works is that you choose a license for your shared data and others must abide by that license in order to use your data. If they do not follow the license, they cannot use the data. Repositories like licenses because they make the terms of use clear, so you'll often see repositories offer one license or a set of licenses for data hosted in the repository. For example, the repository Figshare uses a default Creative Commons Zero (CC0) waiver for all the data on its platform (Figshare 2014b).

Common licenses for data

Two common groups of licenses exist for public data sharing: Creative Commons licenses (www.creativecommons.org) and Open Data Commons licenses (www.opendatacommons.org). Creative Commons licenses actually apply to all kinds of work under copyright, so you're likely to see these licenses on everything from open access articles to images to blog posts. Open Data Commons licenses, on the other hand, were designed to cover data and databases. Where both a Creative Commons and an Open Data Commons license exist with the same conditions, they are roughly equivalent and you can use either one for your data. However, databases should be licensed under an Open Data Common license as these licenses specifically take database rights into account.

Creative Commons and Open Data Commons both offer a suite of licenses to use for content (see Table 10.2). The most basic is the attribution license (CC BY and ODC-By), which allows others to use the licensed content in any way they like, so long as they provide credit to the content creator. Other licenses build on this baseline by adding restrictions, such as share alike. The share-alike licenses (CC BY-SA and ODC-ODbL) require attribution and that any works using or derived from the original be made freely available under a similar share-alike license. This type of share-alike license is common in open source software. Creative Commons also offers a license that prohibits the creation of derivative works (CC BY-ND), meaning you cannot modify the content, and another license forbidding the use of content for commercial purposes (CC BY-NC). Finally, Creative Commons has several combination licenses that stack restrictions like attribution-non commercial-share alike (CC BY-NC-SA) and attribution-non commercial-no derivatives (CC BY-NC-ND).

Apart from licenses that put specific restrictions on the use of copyrighted material, there are two special public domain licenses/waivers: Creative Commons Zero (CC0) and the Public Domain Dedication and License from Open Data Commons (PDDL). These two waivers stipulate that the content has absolutely no restrictions and is in the public domain. Both CC0 and PDDL are useful to both demark content that has no known copyright and to release all known copyright restrictions on a copyrighted work.

Table 10.2 Summary of the Creative Commons and Open Data Commons licenses

License type	Creative Commons license	Open Data Commons license	License description
Public domain	CC0	PDDL	Licensed content is in the public domain with no restrictions on reuse
Attribution	CC BY	ODC-By	All forms of reuse allowed so long as you attribute the content creator
Attribution, share alike	CC BY-SA	ODC-ODbL	Reuse allowed with attribution and the use of an equivalent share-alike license on content derived from the original
Attribution, non-commercial	CC BY-NC	-	Reuse allowed for non-commercial purposes with the requirement of attribution
Attribution, no derivatives	CC BY-ND	-	Original work cannot be modified during reuse and you must attribute
Attribution, non-commercial, share alike	CC BY-NC-SA	-	You must provide attribution, reuse the content for only non-commercial purposes, and place a share-alike license on derivative works
Attribution, non-commercial, no derivatives	CC BY-NC-ND	-	You cannot modify the original or use it commercially and you must provide attribution

Choosing a license

When sharing data publicly, you should opt to use a license over leaving data unlicensed because licenses make reuse rights clear. A license specifically states what people are allowed and what they are not allowed to do with the data. Without a license, others are left trying to figure out if copyright applies to your dataset (a thorny challenge described in Section 10.1.1) and what this means for reuse. The other clarity afforded by a standard license is that they are international. With different restrictions on copyright in different countries, using a standard license is the easiest way to streamline conditions of reuse for anyone in any country.

Due to the fact that copyright does not apply to all datasets, the recommended license for research data is a public domain waiver (either CC0 or PDDL). Data exists in the public domain in the absence of copyright and this waiver reflects that. This recommendation also echoes the Panton Principles, which are a set of recommendations for openly sharing data created by a group of scientists. The Principles state that "data related to published science should be explicitly placed in the public domain" to maximize reuse (Murray-Rust *et al.* 2010). The public domain waiver is slowly becoming the norm for shared data, with many repositories including Figshare (Figshare 2014b), Dryad (Dryad 2014a), BioMed Central (Cochrane 2013), etc. using the CC0 waiver for deposited datasets.

The one limitation of having public domain content is that there cannot be a legal requirement for attribution via citation. For this reason, some people advocate for an attribution license, either CC BY or ODC-By, for shared data. However, given the common absence of copyright it is best to use a public domain waiver and rely on the moral obligation to require citation when you use another person's data (Krier and Strasser 2014). Just as you must cite a paper if you build on its work, so too should you cite a dataset.

The other benefit of using a public domain waiver over an attribution license is that it avoids "attribution stacking". This happens when you combine multiple datasets (which may themselves have been combined from other datasets) with multiple licenses/contracts to perform a wider analysis. At some point, it becomes difficult to follow license terms and attribution requirements for every single dataset or even determine which particular datasets in a database were part of the analysis. The baseline recommendation is therefore to license data using a public domain waiver with the community expectation of citation.

Even if your data falls under copyright, I recommend an open license such as Creative Commons or Open Data Commons for sharing (an open license is one that permits for reuse). The first reason to choose an open license is that your funder or journal's data sharing policy may require it. For example, the Wellcome Trust's data sharing policy states that it "expects all of its funded researchers to maximise the availability of research data with as few restrictions as possible" (Wellcome Trust 2010). While this does not explicitly state a license, it implies that you should use an open license to share your data with minimum restrictions. The other reason to use an open license on copyrighted data is that sharing data with copyright restrictions makes the data available but not usable. Using an open license on all shared data makes clear the reuse permissions in the presence and absence of copyright and maximizes reuse potential.

Finally, it's worth repeating that even when data legally exists in the public domain, it is left to the data owner to determine when and how to share the data. Once you make data publicly available, others may reuse it but you are free to prevent premature reuse by others, such as prior to publishing your article, while the data remains yours.

10.5 GETTING CREDIT FOR SHARED DATA

When sharing data publicly, the expectation is that anytime someone uses your data, they will cite your data (see Chapter 11 for the mechanics of data citation). This leaves us with the happy problem of how to measure credit for shared datasets. In this respect, data sharing is akin to article publication as you get recognition for both "publishing" your data and any citations the data receives.

10.5.1 The basics of getting credit for your data

Just as with articles, you can add shared datasets to your curriculum vitae (CV) and publication list. Data is an important research output that should be promoted alongside your other research products. You can use datasets as evidence of research output for promotion, tenure, job applications, grant applications, etc. in conjunction with more traditional outputs like published articles. Additionally, as funder data sharing policies further develop, documenting shared datasets as an output of previous grants is a good idea for grant renewal.

Beyond recognizing that you made data available, you can also promote any citations your data receives. At present there are few ways to track data citations, though this will change as data sharing and data citation become more common. The current main source for data citation information is Thompson Reuter's Data Citation Index, which is akin to Thompson Reuter's Web of Science but for data instead of articles. Unfortunately, this tool is only available via subscription, most often through an institutional library. Limited data citation tracking is also available via CrossRef. The development of tools for tracking data citation indicates a devotion of resources and that at some point data citation counts will be more readily available.

10.5.2 Altmetrics

A discussion on measuring credit for a newly important type of research output would not be complete without a mention of altmetrics. Altmetrics, short for alternative metrics, are measurements of researcher impact beyond article citation count, journal impact factor, and h-index (Priem *et al.* 2011). So for an article, its alternative metrics could include the number of article views and downloads, the number of article mentions on Twitter, the number of times it is publicly bookmarked or saved to a citation manager, etc. These metrics aren't just for articles, however, they also cover datasets, code, posters, presentations, and other types of research output. Altmetrics give you a picture of your whole research impact in both scholarly and public contexts.

Altmetrics capture success in ways not possible by traditional metrics. For example, in 2012 computer scientist Steve Pettifer was surprised to learn that his 2008 article on bibliographic tools for digital libraries (Hull *et al.* 2008) had over 53,000 downloads. The obvious success of this article was not something Pettifer was capturing in his CV, which only listed the journal's impact factor and the number of article citations – about 80. Happily, he was able to add altmetrics to his CV to better capture this article's impact and successfully apply for promotion (Kwok 2013). The goal of altmetrics is to quantify non-traditional types of impact such as how many people viewed your dataset, how much discussion your presentation generates on Twitter, or the number of times people copy and use your research code, etc. All of these things indicate that you are doing important work, even when this importance is not reflected in your citation count.

Altmetrics are a natural ally for data sharing because shared data is a non-traditional research product. The publishing system is still establishing efficient structures for tracking data citation so altmetrics offer a way to quantify impact in the meantime. Additionally, if you are going to claim credit for a non-traditional research product, it's not that much further to use non-traditional measurements. Both altmetrics and data sharing are fundamentally about getting credit for work beyond the article and its citations.

While altmetrics include a wide range of measurements, there are currently three main sources for compiled altmetrics: ImpactStory (www.impactstory.org), Plum Analytics (www.plumanalytics.com), and Altmetric (www.altmetric.com). If you are interested in looking at your personal metrics, I recommend starting with ImpactStory as they specialize in individual researcher profiles and are advocates of openness.

10.6 CHAPTER SUMMARY

Data sharing is increasingly becoming a standard part of the research process, as scientists share not only published results but also the data that supports these publications. This data sharing exists within an intellectual property framework of copyright, licensing and contracts, and patents. Copyright, which does not always apply to research data, provides the data creator exclusive rights to use the copyrighted work for a finite period of time. If the data falls under copyright, a license or contract will allow other people to use your shared data. Patents, on the other hand, protect inventions supported by data, necessitating that you do not share data prior to patenting your invention.

The most common type of data sharing is that which occurs with your co-workers and your future self. This is the best test of how well you managed the data, as data that is documented, organized, and put into a form supporting longevity will be easier to reuse.

Collaborations are also an important data sharing regime. In this case, data management planning before the start of the project is critical in order to establish conventions that will facilitate data collection and use. You should also plan on discussing data ownership and who will take charge of the data after the project ends. As with local data sharing, good data management practices throughout the project will make it easier to collaborate and share data.

Public data sharing is a new type of data sharing for many scientists, with public research funders enacting data sharing policies since the mid-2000s. Researchers expected to share their data publicly should plan to share any data that supports a publication at the time you publish, unless an applicable data sharing policy says otherwise. Deposit data into a repository, which is a hands-off way to share that supports data longevity. Look to suggestions from your journal, where your peers share, and recommendations from your local data management/library support to choose the best repository for your data.

Once you publicly share data, you can claim it as a research output on your CV, for the purposes of promotion, etc. As with articles, you get credit for both the research product and its citations. Also consider using altmetrics to quantify outside use of your datasets.

11

DATA REUSE AND RESTARTING THE DATA LIFECYCLE

Paleontologists have a problem with dinosaur teeth. They are easy to identify by family but difficult to identify by species, particularly because it's rare to find a complete skull containing teeth. This means that many museum collections have teeth identified as the Tyrannosauridae family or the Dromaeosauridae family with no more specific species information (Farke 2013). In 2013, researchers Larson and Currie decided to do something about this by analyzing over 1,200 dromaeosaurid teeth. Their analysis found that it was possible to differentiate between teeth from species of different time periods because, while it is difficult to distinguish minor features in a handful of teeth, over such a large number of samples the differences between each species' teeth become more apparent (Larson and Currie 2013).

The interesting thing about this study is that Larson and Currie did not have direct access to each of the 1,200 teeth studied. Instead, they relied on a combination of local samples and tooth measurement tables from other researchers to perform such a wide analysis. In short, the research was only possible because other scientists made their data available for reuse. The ability to perform novel research that would not otherwise be possible for lack of data is one of the major benefits of the new availability of shared research data.

As a direct benefit of data sharing, we are opening up the possibilities for what we can study to include broader analysis, meta-analysis, correlating diverse types of data, and using someone else's data as the starting point for your research. All of these types of analysis are possible through greater access to data now that scientific research is heading to a point where it may be as easy to find research data as it is to find published articles. However, with new data sharing systems comes the challenge in actually finding data and the responsibility to properly cite outside datasets. So as we begin the data lifecycle all over again, let us look at the systems for finding data and properly giving credit for data reuse.

11.1 FINDING AND REUSING DATA

As the requirements to share research data are fairly new, we're still developing systems for collecting and discovering research data. Therefore, finding a particular dataset is currently not as easy as finding a published article, though this will improve over time. In the meantime there are several strategies that can help you find and use shared research data in your work.

11.1.1 Finding data

There are a few general approaches you can use for finding data to use in your research. We'll go through these strategies in more detail but I first want to recommend contacting your local research librarian whenever you have trouble finding data. Librarians are specialists in finding information and, while that information has historically been text- and print-based, librarians also have expertise in finding data for research purposes. So don't spend time being frustrated that you can't find a particular dataset because there is often help available.

One way to look for data happens when you read an article and want to see the data that supports it, either to reuse the data or validate the article's findings. Such access is now frequently possible due to the prevalence of data sharing requirements. Note that it does help to know the funding source to see if the data falls under a sharing mandate – this can give you leverage to directly request data if the data is not publicly available.

To find a known dataset, start with the published article itself. Data is sometimes placed in an article's supplementary material or linked to from the article itself. With more journals requiring data sharing, it will become common to find links from a journal article to the corresponding data in an external repository. If the location of the data is not apparent from the article but you know that the data is required to be shared, you have a couple of options. The first is to look into the author's website or CV and see if they list the data as being available somewhere. If this doesn't work, contact the author directly to request access to their data. For those under journal or funder sharing policies (see Chapter 10), they are obligated to provide a copy of their data so long as the data is not sensitive. None of these strategies is foolproof, as older data gets lost and email addresses change (Vines *et al.* 2014), but they can be a successful strategy for getting access to data that corresponds to a known article.

If you are just looking for a general type of data and not data from a specific article, your search strategy will be different. A good place to start a more general search is with a subject-specific index if one exists for your specialty. For example, the Neuroscience Information Framework (NIF 2014) lists a wide range of web-based neuroscience resources and features a search portal to help you find specific neuroscience data. Such indexes don't necessarily collect data but instead point to a number of resources on a particular topic. Indexes and databases may also be available to you via your institution's research library. Not all libraries subscribe to data resources but they are a growing part of library collections so it's worth at least looking for them.

In the absence of an index or library database, consider where your target data might live. Look in data repositories that are popular in your field and refer to the list of data repositories at re3data (www.re3data.org) to see what other disciplinary repositories exist. Also consider outside sources of data like government agencies, research foundations, special interest groups, and other organizations, as they often make information available relating to their work. For example, the National Oceanic and Atmospheric Administration (NOAA) in the United States is an excellent resource for data related to climate. Like with any other type of information, always be sure to evaluate the source of any data you find to be sure it is trustworthy.

Finally, know that as data sharing becomes more routine, it will also become easier to find data for reuse purposes. The research process is currently transitioning to a regime of data sharing which means that many systems for data sharing and reuse are still in development. The goal is for it to someday be as easy to find data from an article as it is to find the article itself.

11.1.2 Data reuse rights

Once you find data you would like to use in your research, you must consider what you are allowed to do with that data. For example, when data is under copyright, most copyright laws allow content to be reused but not republished, with some exceptions. So when using another person's data for your research, you need to be aware of copyright and licensing conditions just as you need to be aware of these concepts when sharing your own data (see Chapter 10).

The easiest type of data reuse comes with data that is openly licensed or placed in the public domain. For publicly shared datasets, that often means data is under a Creative Commons or an Open Data Commons license (see Chapter 10). The benefit of using data under these conditions is that the license will explicitly state what you can and cannot do with the data, making reuse rights clear. Such open licenses, with the occasional exceptions such as limiting commercial research, allow you to both reuse and republish the originally licensed data. Basically, using data under an open license is the best-case scenario because you have relatively broad permissions to do what you need with the data.

Data obtained under a contract can also provide clarity for reuse, though often with more restrictions than for openly licensed content. As contracts usually contain stipulations for what you can and cannot do with the data, it simply becomes a matter of following the contract provisions. Do note that contracts can supersede rights that you would otherwise have under copyright law. For example, if you obtain a collection of factual data (see Chapter 10) under a contract that forbids further sharing of the data, you are not allowed to share the data even though factual data is public domain and thus shareable. Because you agreed to the contract, you must abide by its conditions even when those conditions go against what is otherwise legally allowed.

When you find data that you want to use that is not clearly licensed or listed with reuse permissions, things are more difficult. In this case, you are usually allowed to conduct research on the material but need explicit permission to republish the original data. Remember that copyright largely concerns reproduction, meaning that your limited rights to use copyrighted data do not extend to republication. The major exception to this comes when data consists of facts, such as the locations of the peaks in the infrared spectrum of benzene or the names of all of the species in the Drosophilidae family. Where data consists of facts, you are usually allowed to use and republish those facts assuming that you also cite their source. This means that you can mine facts from a published article, assemble them, and republish them.

The other exception to the ban on republishing data comes under fair use/fair dealing, where applicable under your country's copyright law. Such laws often allow for limited reproduction of a copyrighted work, such as by quoting a short passage of text as you would in an article. In the case of data, this would correspond to reproducing a very small portion of the dataset. In all other cases, you must obtain permission to republish the data from the copyright holder. In the end, you can use data that does not have clear reuse permissions but you must be careful how you present it. For more clarity on this, I recommend SURF's report on the legal status of raw data and, particularly, their brief guide on using someone else's research data (SURF 2014).

11.1.3 Using someone else's data

After finding someone else's data comes the challenge of actually using that data. This can be more or less difficult depending on how well the data was prepared for sharing. In particular, two of the biggest challenges you will encounter are lack of documentation and errors in the data.

Adequate documentation is one of the most important parts of reusing a dataset as you need to understand the details of a dataset in order to use it. For example, you cannot use a dataset if you are unable to determine the meanings of the variable names. So always start by looking for the documentation when reusing a dataset. The best datasets will come with detailed documentation, such as a README file or a data dictionary, though any documentation is better than none. You can refer to the published article in the absence of other documentation, though this rarely provides information on the dataset itself. The best strategy to counter limited documentation is usually to contact the data creator for more information. Hopefully, you will be able to interpret the dataset between the documentation that is available and talking with the original researchers.

Errors in a dataset are another problem you might encounter when using someone else's data. Such errors include inconsistencies, poor null values (see Chapter 6), missing values, and wrong values. Even if you don't find errors in a cursory scan of the data, it is still worth performing some basic tests on the data to check for quality. For example, making a simple graph of the data (as discussed in Chapter 6) is an easy way to check for errors before using a dataset for more complex analyses. The other benefit of performing such quality checks is that you simultaneously gain an understanding of the data. So I encourage you to play around with a dataset before you reuse it to better know how you will have to process the data before analysis.

It's a good idea to reflect on the challenges in reusing someone else's data. This process provides a great opportunity to consider your own data management as nothing teaches you data management quite like having to use someone else's data. If every time you encounter an issue in using someone else's data, you work to improve that issue in your own data management, you will develop good data management habits. While I hope that every dataset you wish to use is easy to understand and of high quality, you can save troubles later if you improve your own habits whenever data reuse frustrations arise.

11.2 CITING DATA

If it's now possible to share and reuse original research data, there needs to be a way to cite someone else's dataset whenever you use it. Just as with articles, researchers should get credit for their shared data so there is an obligation to cite any data that you use. The procedures for citing original research data have only very recently been established as data sharing requirements themselves are so new. Therefore, you may not find this information in style guides or article citation policies until research data citation becomes more common.

11.2.1 Citation format

When you publish an article based on research that uses an outside dataset, you must include the citation for that data as a reference in your article. Data citations should go directly alongside any article citations. In fact, the only difference between citing a dataset and citing an article is the format of the citation itself and not the mechanics of citation. Data is a research product equal to any other research product and the citation should reflect that. This is actually the first principle of the *Joint Declaration of Data Citation Principles* put out by Force11, an international group dedicated to research communication and scholarship (Force11 2013).

The citation itself should include at least the following information (Starr and Gastl 2011; CODATA-ICSTI Task Group on Data Citation Standards and Practices 2013 ; Swoger 2013; California Digital Library 2014):

- Creator
- Publication Year
- Title
- Publisher
- Identifier

All of this citation information, like publication year and the title of the dataset, should be available from the repository hosting the data. The "Creator" is akin to the author of a dataset and can be one person, multiple people, or even an organization. The "Publisher" is the repository that hosts the data. The "Identifier" is the Digital Object Identifier (DOI) or other permanent identifier for the data; if no DOI is available, use the dataset's URL. The identifier is actually one of the most important parts of the citation, as DOI's help with tracking data citations.

Your chosen citation style, like APA or a specific journal style, may have a recommended format for data citations. This is fine to use so long as you include the minimum information listed above. You may also find a formatted citation given in the repository alongside the dataset you're reusing; this citation may or may not contain the necessary citation information (Mooney and Newton 2012). In the absence of a preferred data citation format, use the following format:

> Creator (PublicationYear): Title. Publisher. Identifier

This is the recommended citation format given by DataCite, an international group working to standardize data citation (Starr and Gastl 2011). Here is an example of a data citation in this format:

> Zehr SM, Roach RG, Haring D, Taylor J, Cameron FH, Yoder AD (2014) Data from: Life history profiles for 27 strepsirrhine primate taxa generated using captive data from the Duke Lemur Center. Dryad Digital Repository. http://dx.doi.org/10.5061/dryad.fj974

Your data citation should, at a minimum, include the five major elements already listed but you can also include other information such as (CODATA-ICSTI Task Group on Data Citation Standards and Practices 2013; Swoger 2013; California Digital Library 2014):

- Version
- Series
- Resource Type
- Access Date

Version and series information will not be available for every dataset but are helpful to include in the citation when given. "Resource Type" is nominally "Data set" but can also be "Image", "Sound", "Software", "Database", or "Audiovisual". All of this information is helpful in a data citation but not essential.

A few cases arise where you may want to offer more specifics than the data citation format provides, for example, if you only used a portion of a large dataset or you used a dataset that is continually being added to, as for longitudinal or weather data. In these cases, it's best to use the standard citation format and describe further details within the text that describes your research (Kratz 2013). This allows you to maintain the standard format while still providing additional information.

11.2.2 Other citation considerations

Using the correct citation format is the biggest component of citing a dataset, but there are a few other considerations to be aware of. The first is that simply using someone else's publicly available dataset does not require giving that person author status when you publish that work. This suggestion sometimes appears in articles about sharing datasets (Anon 2014c; Oransky 2014; Roche *et al.* 2014), as the practice is so new. While you would provide author-level credit to a collaborator when you use their data, only a citation is necessary when using publicly available data. The distinction is that you work closely with a co-author throughout the research process whereas using someone's data is akin to building on their published research in your work, which you credit through citation. It is certainly possible that the creator of a reused dataset can become a co-author but the default is citation.

The other thing to note is that sometimes you may be expected to cite the corresponding published article when you cite a dataset. For example, all datasets in the repository Dryad have two listed citations, one for the article and one for the data, with the expectation that you will cite both. However, citing an article whenever you cite a dataset is not strictly required. Datasets do not always correspond to a published article and sometimes you can use a dataset independent of the description of its original research. Nevertheless, it's often the case that reading the article makes you understand the data better, resulting in a citation for both the article and the dataset. In the end you should use your best judgment as to whether you need to cite the article in addition to the data.

11.3 RESTARTING THE DATA LIFECYCLE

Data reuse is the last stage of the data lifecycle (see Chapter 2) but also the beginning of a new cycle. Old data feeds into new projects, with the stages of data collection, data analysis, and disseminating your results occurring once again. Data plays a very important role in this process and its importance is only going to grow as data sharing and reuse becomes more prominent. Therefore, you should start thinking about your data as an important research product that needs to be cared for during and after a research project. It is only through good data management that we can realize research data's full potential.

Now that we've reached the end of both the data lifecycle and the book, let us take a moment to reflect on our journey along the data roadmap from Chapter 2. We've covered everything from data management planning through data organization, documentation, and analysis, taken a detour to review several types of storage and data security, and wrapped up with sharing and reusing data. I fervently hope that you found tools along the way that will help you take better care of your data.

As you continue to explore, research, and incorporate data management strategies into your workflow, remember a few things. First, you don't have to do everything all at once. Data management is the compilation of a lot of small steps that add up to good practices. Try working on one practice at a time until good habits become a routine part of what you do. Any little bit you do helps to take care of your data better.

Data management also gets easier with time. This is partly because you will learn which practices fit into your research workflows and how to adapt other strategies for your data. Good data habits will also become a routine part of doing research. The goal is to reach a point where good data management simply becomes a background part of doing your research.

When dealing with data becomes frustrating, use it as an opportunity to manage your data better so the frustrations don't happen again. For example, if you are having problems understanding a co-worker's data from her notes, think about how you can improve your own notes to make your data clearer to a future co-worker or your future self. Too often, dealing with data is frustrating but you do have the power to eliminate those frustrations through reflection and conscious data management.

Finally, remember that data management is a living process. It is not some goal you reach and then you never have to do data management again. Managing your data well requires effort but that effort pays dividends later when you do not spend hours trying to find, understand, or reuse your data. It's my hope that by doing conscious and continual data management, you can significantly reduce the everyday frustrations of dealing with research data. Your data should work for you, not against you because scientific research on its own is challenging enough.

REFERENCES

3TU. Datacentrum, 2013. *3TU. Datacentrum*. Available at: http://datacentrum.3tu.nl/en/home/ [accessed October 26, 2014].

AIP, 2014. *PACS 2010 Regular Edition*. Available at: http://www.aip.org/pacs [accessed December 28, 2013].

Akers, K., 2014. A growing list of data journals. *Data@MLibrary*. Available at: http://mlibrarydata.wordpress.com/2014/05/09/data-journals/ [accessed October 26, 2014].

Alexander, R., 2013. Reinhart, Rogoff ... and Herndon: the student who caught out the profs. *BBC News*. Available at: http://www.bbc.com/news/magazine-22223190 [accessed October 14, 2014].

Anderson, N., 2009. "Anonymized" data really isn't – and here's why not. *Ars Technica*. Available at: http://arstechnica.com/tech-policy/2009/09/your-secrets-live-online-in-databases-of-ruin/ [accessed August 11, 2014].

Anon, 1980. *Patent and Trademark Law Amendments Act*. 96th United States Congress.

Anon, 1991. *Feist Publications, Inc. v. Rural Telephone Service Company, Inc.* US Supreme Court.

Anon, 1997. *Copyright and Rights in Databases Regulations 1997*. United Kingdom.

Anon, 2014a. Data loss statistics. *Boston Computing Network*. Available at: http://www.bostoncomputing.net/consultation/databackup/statistics/ [accessed December 2, 2014].

Anon, 2014b. STAP retracted. *Nature*, 511(7507), pp. 5–6.

Anon, 2014c. Share alike. *Nature*, 507(7491), pp. 140. Available at: http://www.nature.com/news/share-alike-1.14850 [accessed November 3, 2014].

Association for Computing Machinery, 2014. *ACM Computing Classification System*. Available at: http://www.acm.org/about/class [accessed December 28, 2013].

Barbaro, M. and Zeller, T., 2006. A face is exposed for AOL Searcher No. 4417749. *New York Times*. Available at: http://www.nytimes.com/2006/08/09/technology/09aol.html?pagewanted=all&_r=0 [accessed August 12, 2014].

Barber, C.R., 2011. Yankaskas settles appeal, agrees to retire from UNC. *The Daily Tar Heel*. Available at: http://www.dailytarheel.com/index.php/article/2011/04/yankaskas_settles_appeal_agrees_to_retire_from_unc [accessed July 6, 2014].

Beach, B., 2013. How long do disk drives last? *Backblaze Blog*. Available at: http://blog.backblaze.com/2013/11/12/how-long-do-disk-drives-last/ [accessed May 20, 2014].

Bhattacharjee, Y., 2013. Diederik Stapel's audacious academic fraud. *The New York Times*. Available at: http://www.nytimes.com/2013/04/28/magazine/diederik-stapels-audacious-academic-fraud.html?pagewanted=all [accessed September 14, 2014].

Biotechnology and Biological Sciences Research Council, 2010. *Data Sharing Policy*. Available at: http://www.bbsrc.ac.uk/web/FILES/Policies/data-sharing-policy.pdf [accessed February 2, 2014].

Bloom, T., 2013. Data access for the open access literature: PLOS's data policy. *PLOS ONE*. Available at: http://www.plos.org/data-access-for-the-open-access-literature-ploss-data-policy/ [accessed January 26, 2014].

Borgman, C.L., 2012. The conundrum of sharing research data. *Journal of the American Society for Information Science and Technology*, 63(6), pp. 1059–1078. Available at: http://onlinelibrary.wiley.com/doi/10.1002/asi.22634/full [accessed March 10, 2015].

Brown, C., 2010. My data management plan – a satire. *Living in an Ivory Basement*. Available at: http://ivory.idyll.org/blog/data-management.html [accessed July 27, 2014].

Bruna, E.M., 2014. The opportunity cost of my #OpenScience was 36 hours + $690. *The Bruna Lab*. Available at: http://brunalab.org/blog/2014/09/04/the-opportunity-cost-of-my-openscience-was-35-hours-690/ [accessed October 11, 2014].

Buckheit, J.B., and Donoho, D.L., 1995. WaveLab and reproducible research. In *Lecture Notes in Statistics Volume 103* (pp. 55–81). New York: Springer.

Cajochen, C., Altanay-Ekici, S., Münch, M., Frey, S., Knoblauch, V., and Wirz-Justice, A., 2013. Evidence that the lunar cycle influences human sleep. *Current Biology: CB*, 23(15), pp. 1485–1488. Available at: http://www.sciencedirect.com/science/article/pii/S0960982213007549 [accessed March 10, 2015].

California Digital Library, 2014. Data citation. *Data Pub*. Available at: http://datapub.cdlib.org/datacitation/ [accessed November 2, 2014].

Callaghan, S., 2013. A list of data journals. *PREPARDE Project*. Available at: http://proj.badc.rl.ac.uk/preparde/blog/DataJournalsList [accessed October 26, 2014].

Chacon, S., 2009. *Pro Git*. New York City: Apress.

Check Hayden, E., 2013. Privacy loophole found in genetic databases. *Nature News & Comment*. Available at: http://www.nature.com/news/privacy-loophole-found-in-genetic-databases-1.12237 [accessed September 21, 2014].

Chue Hong, N., 2013. Which journals should I publish my software in? *Software Sustainability Institute*. Available at: http://www.software.ac.uk/resources/guides/which-journals-should-i-publish-my-software [accessed April 8, 2014].

Cochrane, J., 2013. Open by default: making open data truly open. *BioMed Central blog*. Available at: http://blogs.biomedcentral.com/bmcblog/2013/08/21/open-by-default-making-open-data-truly-open/ [accessed October 12, 2014].

CODATA-ICSTI Task Group on Data Citation Standards and Practices, 2013. Out of cite, out of mind: the current state of practice, policy, and technology for the citation of data. *Data Science Journal*, 12 (2013), pp. CIDCR1–CIDCR75. Available at: https://www.jstage.jst.go.jp/article/dsj/12/0/12_OSOM13-043/_article [accessed November 2, 2014].

Cohen, J., 2011. Intellectual property. Dispute over lab notebooks lands researcher in jail. *Science (New York, NY)*, 334(6060), pp. 1189–1190. Available at: http://www.ncbi.nlm.nih.gov/pubmed/22144589 [accessed January 26, 2014].

Corbyn, Z., 2010. UEA mishandled "Climategate" Freedom of Information requests, say MPs. *Times Higher Education*. Available at: http://www.timeshighereducation.co.uk/news/uea-mishandled-climategate-freedom-of-information-requests-say-mps/411053.article [accessed October 19, 2014].

Corti, L., 2014. *Managing and Sharing Research Data: A Guide to Good Practice*. Los Angeles: SAGE Publications.

Creative Commons, 2014. *Creative Commons*. Available at: http://creativecommons.org/ [accessed January 28, 2014].

Čurn, J., 2014. A cautionary tale: how a bug in dropbox permanently deleted 8,000 of my photos. *PetaPixel*. Available at: http://petapixel.com/2014/07/31/cautionary-tale-bug-dropbox-permanently-deleted-8000-photos/ [accessed September 23, 2014].

Darwin Core Task Group, 2014. *Darwin Core*. Available at: http://rs.tdwg.org/dwc/ [accessed January 4, 2014].

Daylight Chemical Information Systems Inc., 2014. *SMILES*. Available at: http://www.daylight.com/smiles/index.html [accessed December 3, 2013].

Digital Curation Centre, 2014a. *DMPonline*. Available at: https://dmponline.dcc.ac.uk/ [accessed February 7, 2014].

Digital Curation Centre, 2014b. *Overview of Funders' Data Policies*. Available at: http://www.dcc.ac.uk/resources/policy-and-legal/overview-funders-data-policies [accessed January 28, 2014].

Digital Curation Centre, 2014c. *Disciplinary Metadata*. Available at: http://www.dcc.ac.uk/resources/metadata-standards [accessed January 4, 2014].

Digital Preservation Coalition, 2014. *Media and Formats – Media*. Available at: http://www.dpconline.org/advice/preservationhandbook/media-and-formats/media [accessed May 20, 2014].

Dryad, 2014a. *Frequently Asked Questions*. Available at: http://datadryad.org/pages/journalLookup [accessed July 8, 2014].

Dryad, 2014b. *Integrated Journals*. Available at: http://datadryad.org/pages/integrated-Journals [accessed October 11, 2014].

Dryad, 2014c. *Joint Data Archiving Policy*. Available at: http://datadryad.org/pages/jdap [accessed October 14, 2014].

Dublin Core Metadata Initiative, 2014. *Dublin Core Metadata Initiative*. Available at: http://dublincore.org/ [accessed December 27, 2013].

Economics and Social Research Council, 2012. *Framework for Research Ethics*. Available at: http://www.esrc.ac.uk/_images/framework-for-research-ethics-09-12_tcm8-4586.pdf [accessed August 23, 2014].

Electronic Privacy Information Center, 2014. *Re-identification*. Available at: http://epic.org/privacy/reidentification/#process [accessed August 12, 2014].

El Emam, K., 2013. *Guide to the De-identification of Personal Health Information*. Boca Raton: Taylor & Francis.

El Emam, K., 2014. *Anonymizing Health Data Case Studies and Methods to Get You Started*. Sebastopol CA: O'Reilly Media.

EMC, 2011. *Digital Universe*. Available at: http://www.emc.com/leadership/programs/

digital-universe.htm [accessed December 7, 2014].

Engineering and Physical Sciences Research Council, 2013. *Policy Framework on Research Data – Expectations*. Available at: http://www.epsrc.ac.uk/about/standards/research-data/expectations/ [accessed March 10, 2015].

Engineering and Physical Sciences Research Council, 2014. *Expectations*. Available at: http://www.epsrc.ac.uk/about/standards/researchdata/expectations/ [accessed July 13, 2014].

Enserink, M. and Malakoff, D., 2012. Will flu papers lead to new research oversight? *Science (New York, NY)*, 335(6064), pp. 20, 22.

ESRI, 2014. *GIS Dictionary*. Available at: http://support.esri.com/en/knowledgebase/Gisdictionary/browse [accessed December 28, 2013].

European Parliament, 1995. *EUR-Lex - 31995L0046 - EN - EUR-Lex*. Available at: http://eur-lex.europa.eu/LexUriServ/LexUriServ.do?uri=CELEX:31995L0046:en:HTML [accessed March 10, 2015].

Fanelli, D., 2013. Why growing retractions are (mostly) a good sign. *PLOS Medicine*, 10(12), p. e1001563.

Farke, A., 2013. And this is why we should always provide our data… *The Integrative Paleontologists*. Available at: http://blogs.plos.org/paleo/2013/01/25/and-this-is-why-we-should-always-provide-our-data/ [accessed November 18, 2014].

Favre, H.A., Powell, W.H., and IUPAC, 2014. *Nomenclature of Organic Chemistry: IUPAC Recommendations and Preferred Names 2013*, Cambridge, England: Royal Society of Chemistry.

Ferreri, E., 2011. Breach costly for researcher, UNC-CH. *News Observer*. Available at: http://web.archive.org/web/20110613075205/http://www.newsobserver.com/2011/05/09/1185493/breach-costly-for-researcher-unc.html [accessed July 6, 2014].

Figshare, 2014a. The stuff of nightmares: imagine losing all your research data. *Figshare blog*. Available at: http://figshare.com/blog/The_stuff_of_nightmares_imagine_losing_all_your_research_data/121 [accessed September 23, 2014].

Figshare, 2014b. *How is Your Uploaded Data Licensed?* Available at: http://figshare.com/licensing [accessed October 11, 2014].

Force11, 2013. *Joint Declaration of Data Citation Principles*. Available at: https://www.force11.org/datacitation [accessed November 2, 2014].

Gene Ontology Consortium, 2013. *The Gene Ontology*. Available at: http://www.geneontology.org/ [accessed December 27, 2013].

GitHub, 2014. *Code School – Try Git*. Available at: http://try.github.io/levels/1/challenges/1 [accessed April 22, 2014].

Goben, A. and Salo, D., 2013. Federal research: data requirements set to change. *College and Research Libraries News*, 74(8), pp. 421–425.

Google, 2014. *Google Terms of Service – Privacy & Terms*. Available at: http://www.google.com/policies/terms/ [accessed April 29, 2014].

Gray, N., 2012. *Data is a Singular Noun*. Available at: http://purl.org/nxg/note/singular-data [accessed January 6, 2015].

Harvard University, 2011. Retention of research data and materials. *Harvard University Office of Sponsored Programs*. Available at: http://osp.fad.harvard.edu/content/

retention-of-research-data-and-materials [accessed January 28, 2014].

Herald, P., 2012. Stolen laptop had dissertation stored on it which is due in tomorrow. *The Herald*. Available at: http://www.plymouthherald.co.uk/Stolen-laptop-dissertation-stored-tomorrow/story-16031933-detail/story.html [accessed January 26, 2014].

Holdren, J., 2013. Increasing access to the results of federally funded scientific research. *Office of Science and Technology Policy*. Available at: http://www.whitehouse.gov/sites/default/files/microsites/ostp/ostp_public_access_memo_2013.pdf [accessed November 25, 2013].

Hull, D., Pettifer, S.R., and Kell, D.B., 2008. Defrosting the digital library: bibliographic tools for the next generation. Web. J. McEntyre, ed. *PLOS Computational Biology*, 4(10), p. e1000204.

ICO, 2014. Anonymisation code of practice. *Information Commissioner's Office*. Available at: http://ico.org.uk/for_organisations/data_protection/topic_guides/anonymisation [accessed August 24, 2014].

InChI Trust, 2013. *Find Out About InChI*. Available at: http://www.inchi-trust.org/ [accessed December 3, 2013].

International Health Terminology Standards Development Organisation, 2014. *SNOMED CT*. Available at: http://www.ihtsdo.org/snomed-ct/ [accessed December 27, 2013].

International Organization for Standardization, 1988. *Data Elements and Interchange Formats: Information Interchange – Representation of Dates and Times*, Geneva: International Organization for Standardization.

International Organization for Standardization, 2003. *Geographic Information, Metadata = Information Géographique, Métadonnées*, Geneva: International Organization for Standardization.

International Organization for Standardization, 2007. *Geographic Information – metadata – XML schema implementation = Information géographique – métadonnées – implémentation de schémas XML*. Geneva: International Organization for Standardization.

International Union of Crystallography, 2014. *Crystallographic Information Framework*. Available at: http://www.iucr.org/resources/cif [accessed January 4, 2014].

Iorns, E., 2013. Reproducibility initiative receives $1.3M grant to validate 50 landmark cancer studies. *Science Exchange*. Available at: http://blog.scienceexchange.com/2013/10/reproducibility-initiative-receives-1-3m-grant-to-validate-50-landmark-cancer-studies/ [accessed September 14, 2014].

IUPAC, 2014. *IUPAC Nomenclature Home Page*. Available at: http://www.chem.qmul.ac.uk/iupac/ [accessed December 27, 2013].

Johnston, C., 2011. Ask Ars: how can I securely erase the data from my SSD drive? *Ars Technica*. Available at: http://arstechnica.com/security/2011/03/ask-ars-how-can-i-safely-erase-the-data-from-my-ssd-drive/ [accessed July 21, 2014].

JSON.org, 2014. *Introducing JSON*. Available at: http://www.json.org/ [accessed January 5, 2014].

Kanare, H.M., 1985. *Writing the Laboratory Notebook*. Washington DC: American Chemical Society.

Katz, L., 2011. Stolen laptop contains cancer cure data. *CNET*. Available at: http://www.cnet.com/news/stolen-laptop-contains-cancer-cure-data/ [accessed September 23, 2014].

Keim, B., 2012. Researchers hit pause on controversial killer flu research. *Ars Technica*. Available at: http://arstechnica.com/science/2012/01/researchers-hit-pause-on-controversial-killer-flu-research/ [accessed August 23, 2014].

Kolowich, S., 2011. Security hacks. *Inside Higher Ed*. Available at: http://www.insidehighered.com/news/2011/01/27/unc_case_highlights_debate_about_data_security_and_accountability_for_hacks [accessed July 6, 2014].

Kolowich, S., 2014. Hazards of the cloud: data-storage service's crash sets back researchers. *The Chronicle of Higher Education*. Available at: http://chronicle.com/blogs/wiredcampus/hazards-of-the-cloud-data-storage-services-crash-sets-back-researchers/52571 [accessed May 20, 2014].

Kratz, J., 2013. Data citation developments. *Data Pub*. Available at: http://datapub.cdlib.org/2013/10/11/data-citation-developments/ [accessed November 3, 2014].

Krier, L. and Strasser, C., 2014. *Data Management for Libraries: A LITA Guide*, Chicago: ALA TechSource.

Kwok, R., 2013. Research impact: altmetrics make their mark. *Nature*, 500(7463), pp. 491–493. Available at: http://www.nature.com/naturejobs/science/articles/10.1038/nj7463-491a [accessed October 12, 2014].

Larson, D.W. and Currie, P.J., 2013. Multivariate analyses of small theropod dinosaur teeth and implications for paleoecological turnover through time. *PLOS ONE*, 8(1), p. e54329.

Lawrence Livermore National Laboratory, 2014. *CF Conventions Home Page*. Available at: http://cfconventions.org/ [accessed July 20, 2014].

Lawson, K., 2013. Resources for learning Git and GitHub. *The Chronicle of Higher Education*. Available at: http://chronicle.com/blogs/profhacker/resources-for-learning-git-and-github/48285 [accessed April 22, 2014].

Leopando, J., 2013. World backup day: the 3-2-1 rule. *Security Intelligence Blog*. Available at: http://blog.trendmicro.com/trendlabs-security-intelligence/world-backup-day-the-3-2-1-rule/ [accessed September 23, 2014].

LeVeque, R.J., Mitchell, I.M., and Stodden, V., 2012. Reproducible research for scientific computing: tools and strategies for changing the culture. *Computing in Science and Engineering*, 14(4), pp. 13–17.

Levine, M., 2013. Facts and data. *University of Michigan Library*. Available at: http://www.lib.umich.edu/copyright/facts-and-data [accessed October 21, 2014].

Levkina, M., 2014. How to follow the 3-2-1 backup rule with Veeam Backup & Replication. *Veeam Blog*. Available at: http://www.veeam.com/blog/how-to-follow-the-3-2-1-backup-rule-with-veeam-backup-replication.html [accessed September 23, 2014].

LOCKSS, 2014. *Lots of Copies Keep Stuff Safe*. Available at: http://www.lockss.org/ [accessed May 20, 2014].

Loeliger, J. and McCullough, M., 2012. *Version Control with Git*, 2nd edn, Sebastopol, CA: O'Reilly Media.

LOIRP, 2014. *Moonviews*. Available at: http://www.moonviews.com/ [accessed November 17, 2014].

Lunshof, J.E., Chadwick, R., Vorhaus, D.B., and Church, G.M., 2008. From genetic privacy to open consent. *Nature Reviews. Genetics*, 9(5), pp. 406–411. Available at: http://dx.doi.

org/10.1038/nrg2360 [accessed August 23, 2014].

Macey, R., 2006. One giant blunder for mankind: how NASA lost moon pictures. *The Sydney Morning Herald*. Available at: http://www.smh.com.au/news/science/one-giant-blunder-for-mankind-how-nasa-lost-moon-pictures/2006/08/04/1154198328978.html [accessed November 11, 2014].

Marcus, A., 2012. Journal retracts protein paper from scientist who misused deceased mentor's data. *Retraction Watch*. Available at: http://retractionwatch.com/2012/05/17/journal-retracts-protein-paper-from-serial-data-thief-who-used-deceased-mentors-name/ [accessed January 26, 2014].

Marcus, A., 2013a. Dispute over data forces retraction of wasp paper. *Retraction Watch*. Available at: http://retractionwatch.com/2013/06/20/dispute-over-data-forces-retraction-of-wasp-paper/ [accessed January 26, 2014].

Marcus, A., 2013b. Med student loses paper when former boss claims right to data. *Retraction Watch*. Available at: http://retractionwatch.com/2013/02/22/med-student-loses-paper-when-former-boss-claims-right-to-data/ [accessed January 26, 2014].

Marcus, A., 2013c. Not our problem: journal bows out of data dispute after U Minn challenges previous statement. *Retraction Watch*. Available at: http://retractionwatch.com/2013/07/31/not-our-problem-journal-bows-out-of-data-dispute-after-u-minn-challenges-previous-statement/ [accessed January 26, 2014].

Marcus, A., 2013d. Our bad! Researchers take colleagues' data, lose paper. *Retraction Watch*. Available at: http://retractionwatch.com/2013/07/05/our-bad-researchers-take-colleagues-data-lose-paper/ [accessed January 26, 2014].

Marcus, A., 2013e. Influential Reinhart–Rogoff economics paper suffers spreadsheet error. *Retraction Watch*. Available at: http://retractionwatch.wordpress.com/2013/04/18/influential-reinhart-rogoff-economics-paper-suffers-database-error/ [accessed October 14, 2014].

Marcus, A., 2014. Doing the right thing: authors retract brain paper with "systematic human error in coding". *Retraction Watch*. Available at: http://retractionwatch.com/2014/01/07/doing-the-right-thing-authors-retract-brain-paper-with-systematic-human-error-in-coding/ [accessed January 26, 2014].

Marshall, E., 2001. Bermuda rules: community spirit, with teeth. *Science*, 291(5507), pp. 1192.

Mathôt, S., 2014. Git for non-hackers pt. 1: organizing your research one commit at a time. *COGSCIdotNL*. Available at: http://www.cogsci.nl/blog/miscellaneous/228-git-for-non-hackers-pt-1-organizing-your-research-one-commit-at-a-time [accessed April 22, 2014].

Mearian, L., 2011. World's data will grow by 50X in next decade, IDC study predicts. *Computer World*. Available at: http://www.computerworld.com/article/2509588/data-center/world-s-data-will-grow-by-50x-in-next-decade--idc-study-predicts.html [accessed December 7, 2014].

Medical Research Council, 2014. *Section 2: Guidelines and Standards*. Available at: http://www.mrc.ac.uk/research/research-policy-ethics/good-research-practice/guidelines-and-standards/ [accessed July 14, 2014].

Mobley, A., Linder, S.K., Braeuer, R., Ellis, L.M., and Zwelling, L., 2013. A survey on data reproducibility in cancer research provides insights into our limited ability to translate findings from the laboratory to the clinic. *PLOS ONE*, 8(5), p. e63221.

Mooney, H. and Newton, M., 2012. The anatomy of a data citation: discovery, reuse, and credit. *Journal of Librarianship and Scholarly Communication*, 1(1), p. eP1035.

Munroe, R., 2013. *xkcd: ISO 8601*. Available at: http://xkcd.com/1179/ [accessed December 28, 2013].

Murray-Rust, P., Neylon, C., Pollock, R., and Wilbanks, J., 2010. *Panton Principles*. Available at: http://pantonprinciples.org/ [accessed October 12, 2014].

Narayanan, A., 2014. No silver bullet: de-identification still doesn't work. *Freedom to Tinker*. Available at: https://freedom-to-tinker.com/blog/randomwalker/no-silver-bullet-de-identification-still-doesnt-work/ [accessed September 21, 2014].

NASA, 2009. *The Apollo 11 Telemetry Data Recordings: A Final Report*. Available at: http://www.honeysucklecreek.net/Apollo_11/tapes/398311main_Apollo_11_Report.pdf [accessed November 11, 2014].

NASA, 2013a. *FITS Standard Page*. Available at: http://fits.gsfc.nasa.gov/fits_standard.html [accessed January 4, 2014].

NASA, 2013b. *NASA Scientific and Technical Information Program*. Available at: http://www.sti.nasa.gov/sti-tools/#thesaurus [accessed December 27, 2013].

NASA, 2014. *DIF Writer's Guide*. Available at: http://gcmd.gsfc.nasa.gov/add/difguide/index.html [accessed January 4, 2014].

National Center for Biotechnology Information, 2013. *NCBI Taxonomy*. Available at: http://www.ncbi.nlm.nih.gov/Taxonomy/taxonomyhome.html/ [accessed December 27, 2013].

National Digital Information Infrastructure and Preservation Program, 2013. *Introduction to Digital Formats for Library of Congress Collections*. Available at: http://www.digitalpreservation.gov/formats/intro/intro.shtml [accessed July 15, 2014].

National Health and Medical Research Council, The Australian Research Council, and Universities Australia, 2007. *Australian Code for the Responsible Conduct of Research*. Available at: http://www.nhmrc.gov.au/_files_nhmrc/publications/attachments/r39.pdf [accessed July 13, 2014].

National Institutes of Health, 2003. *Final NIH Statement on Sharing Research Data*. Available at: http://grants.nih.gov/grants/guide/notice-files/NOT-OD-03-032.html [accessed October 19, 2014].

National Institutes of Health, 2012. *NIH Grants Policy Statement*. Available at: http://grants.nih.gov/grants/policy/nihgps_2012/nihgps_ch8.htm [accessed January 28, 2014].

National Library of Medicine, 2013. *Medical Subject Headings*. Available at: http://www.nlm.nih.gov/mesh/ [accessed December 27, 2013].

National Science Foundation, 2013. *National Science Foundation Grant General Conditions*. Available at: http://www.nsf.gov/pubs/policydocs/gc1/jan13.pdf [accessed February 2, 2014].

Nature Publishing Group, 2006. *Availability of Data and Materials*. Available at: http://www.nature.com/authors/policies/availability.html [accessed February 18, 2014].

NCBI, 2011. *Adding Value to your Submission*. Available at: http://www.ncbi.nlm.nih.gov/books/NBK53714/ [accessed October 12, 2014].

NIF, 2014. *Neuroscience Information Framework*. Available at: http://www.neuinfo.org/ [accessed November 23, 2014].

NPR, 1996. *Disputes Rise Over Intellectual Property Rights*. Available at: http://www.cptech.org/ip/npr.txt [accessed January 26, 2014].

NSF, 2013. *Dissemination and Sharing of Research Results*. Available at: http://www.nsf.gov/bfa/dias/policy/dmp.jsp.

NSF Division of Earth Sciences, 2010. *Dissemination and Sharing of Research Results*. Available at: http://www.nsf.gov/geo/ear/2010EAR_data_policy_9_28_10.pdf [accessed October 11, 2014].

NYU Health Sciences Library, 2014. *How to Avoid a Data Management Nightmare*. Available at: https://www.youtube.com/watch?v=nNBiCcBlwRA [accessed April 13, 2014].

OBO Foundry, 2014. *The Open Biological and Biomedical Ontologies*. Available at: http://www.obofoundry.org/ [accessed December 27, 2013].

Office of Research Integrity, 2014. *Data Management, Retention of Data*. Available at: https://ori.hhs.gov/education/products/rcradmin/topics/data/tutorial_11.shtml [accessed July 14, 2014].

Office of the Australian Information Commissioner, 2014. *Privacy Fact Sheet 17: Australian Privacy Principles*. Available at: http://www.oaic.gov.au/privacy/privacy-resources/privacy-fact-sheets/other/privacy-fact-sheet-17-australian-privacy-principles [accessed July 26, 2014].

O'Neal, J.E., 2009. Search for missing recordings ends. *TVTechnology*. Available at: http://www.tvtechnology.com/feature-box/0124/search-for-missing-recordings-ends/202982 [accessed November 11, 2014].

Open Data Commons, 2014. *Open Data Commons*. Available at: http://opendatacommons.org/ [accessed January 28, 2014].

Oransky, I., 2010. Update on Ahluwalia fraud case: researcher faked results, probably committed sabotage, says UCL. *Retraction Watch*. Available at: http://retractionwatch.com/2010/11/26/update-on-ahluwalia-fraud-case-researcher-faked-results-probably-committed-sabotage-says-ucl/ [accessed September 21, 2014].

Oransky, I., 2011. Exclusive: researcher found guilty of misconduct at UCL had been dismissed from Cambridge for data fabrication. *Retraction Watch*. Available at: http://retractionwatch.com/2011/02/08/exclusive-researcher-found-guilty-of-misconduct-at-ucl-had-been-dismissed-from-cambridge-for-data-fabrication/ [accessed September 18, 2014].

Oransky, I., 2012. PLOS ONE GMO cassava paper retracted after data "could not be found". *Retraction Watch*. Available at: http://retractionwatch.com/2012/09/14/plos-one-gmo-cassava-paper-retracted-after-data-could-not-be-found/ [accessed January 26, 2014].

Oransky, I., 2013a. Measure by measure: Diederik Stapel count rises again, to 54. *Retraction Watch*. Available at: http://retractionwatch.com/2013/08/02/measure-by-measure-diederik-stapel-count-rises-again-to-54/ [accessed September 14, 2014].

Oransky, I., 2013b. A real shame: psychology paper retracted when data behind problematic findings disappear. *Retraction Watch*. Available at: http://retractionwatch.com/2013/08/15/a-real-shame-psychology-paper-retracted-when-data-behind-problematic-findings-disappear/ [accessed January 26, 2014].

Oransky, I., 2013c. JCI paper retracted for duplicated panels after authors can't provide original data. *Retraction Watch*. Available at: http://retractionwatch.com/2013/07/19/jci-paper-retracted-for-duplicated-panels-after-authors-cant-provide-original-data/ [accessed January 26, 2014].

Oransky, I., 2013d. NEJM paper on sleep apnea retracted when original data can't be found. *Retraction Watch*. Available at: http://retractionwatch.com/2013/10/30/nejm-paper-on-sleep-apnea-retracted-when-original-data-cant-be-found/ [accessed January 26, 2014].

Oransky, I., 2013e. Spat over tuberculosis study data leads to Expression of Concern. *Retraction Watch*. Available at: http://retractionwatch.com/2013/09/26/spat-over-tuberculosis-study-data-leads-to-expression-of-concern/ [accessed January 26, 2014].

Oransky, I., 2013f. "Unfinished business": Diederik Stapel retraction count rises to 53. *Retraction Watch*. Available at: http://retractionwatch.wordpress.com/2013/04/30/unfinished-business-diederik-stapel-retraction-count-rises-to-54/ [accessed September 14, 2014].

Oransky, I., 2013g. A real shame: psychology paper retracted when data behind problematic findings disappear. *Retraction Watch*. Available at: http://retractionwatch.com/2013/08/15/a-real-shame-psychology-paper-retracted-when-data-behind-problematic-findings-disappear/ [accessed January 26, 2014].

Oransky, I., 2014. Oxford group reverses authorship requirements for sharing data after questions from Retraction Watch. *Retraction Watch*. Available at: http://retractionwatch.com/2014/10/06/oxford-group-reverses-authorship-requirements-for-sharing-data-after-questions-from-retraction-watch/ [accessed December 3, 2014].

Organisation for Economic Co-operation and Development, 2007. *OECD Principles and Guidelines for Access to Research Data from Public Funding*. Available at: http://www.oecd.org/science/sci-tech/38500813.pdf [accessed November 17, 2014].

Pandurangan, V., 2014. On taxis and rainbows. *Medium*. Available at: https://medium.com/@vijayp/of-taxis-and-rainbows-f6bc289679a1 [accessed August 12, 2014].

Pattinson, D., 2013. PLOS ONE launches reproducibility initiative – EveryONE. *PLOS Blog*. Available at: http://blogs.plos.org/everyone/2012/08/14/plos-one-launches-reproducibility-initiative/ [accessed September 14, 2014].

Pearlman, R.Z., 2009. NASA erased first moonwalk tapes, but restores copies. *Space.com*. Available at: http://www.space.com/6994-nasa-erased-moonwalk-tapes-restores-copies.html [accessed November 11, 2014].

Pinheiro, E., Weber, W.-D., and Barroso, L.A., 2007. Failure trends in a large disk drive population. *5th USENIX Conference on File and Storage Technologies*. Available at: http://static.googleusercontent.com/media/research.google.com/en/us/archive/disk_failures.pdf [accessed May 21, 2014].

Pittman, 2012. Insubordination in the lab. *BioTechniques*. Available at: http://www.biotechniques.com/news/Insubordination-in-the-Lab/biotechniques-330913.html [accessed January 26, 2014].

Piwowar, H.A. and Vision, T.J., 2013. Data reuse and the open data citation advantage. *PeerJ*, 1, p. e175.

Priem, J., Taraborelli, D., Groth, P., and Neylon, C., 2011. Altmetrics: a manifesto. *altmetrics.org*. Available at: http://altmetrics.org/manifesto/ [accessed November 26, 2014].

Purdue University, 2013. *Purdue University Research Repository*. Available at: https://purr.purdue.edu/ [accessed October 26, 2014].

Purrington, C., 2011. *Maintaining a Laboratory Notebook*. Available at: http://colinpurrington.com/tips/lab-notebooks [accessed December 7, 2014].

Read, M., 2010. Grad student's thesis, dreams on stolen laptop. *Gawker*. Available at: http://gawker.com/5625139/grad-students-thesis-dreams-on-stolen-laptop [accessed January 26, 2014].

Reinhart, C.M. and Rogoff, K.S., 2010. Growth in a time of debt. *American Economic Review*, 100(2), pp. 573–578. Available at: https://www.aeaweb.org/articles.php?doi=10.1257/aer.100.2.573 [accessed October 14, 2014].

Research Councils UK, 2011. *RCUK Common Principles on Data Policy*. Available at: http://www.rcuk.ac.uk/research/datapolicy/ [accessed October 19, 2014].

Roche, D.G., Lanfear, R., Binning, S.A., Haff, T.M., Schwanz, L.E., Cain, K.E., Kokko, H., Jennions, M.D., Kruuk, L.E.B., 2014. Troubleshooting public data archiving: suggestions to increase participation. *PLOS Biology*, 12(1), p. e1001779.

Royal Society of Chemistry, 2014. *ChemSpider*. Available at: http://www.chemspider.com/ [accessed December 27, 2013].

Royal Society Publishing, 2014. *Philosophical Transactions of the Royal Society of London*. Available at: http://rstl.royalsocietypublishing.org/ [accessed December 27, 2014].

Salo, D., 2013. *Data Storage, Short Term*. Available at: http://www.graduateschool.uwm.edu/forms-and-downloads/researchers/data-management-plan/5DataStorageShort-Term.pdf [accessed March 10, 2015].

Sawyer, K., 1999. Mystery of orbiter crash solved. *Washington Post*. Available at: http://www.washingtonpost.com/wp-srv/national/longterm/space/stories/orbiter100199.htm [accessed January 6, 2015].

Schroeder, B. and Gibson, G.A., 2007. Disk failures in the real world: what does an MTTF of 1,000,000 hours mean to you? *5th USENIX Conference on File and Storage Technologies*. Available at: http://static.usenix.org/events/fast07/tech/schroeder.html [accessed May 21, 2014].

Schwartz, C., 2012. How research into chronic fatigue syndrome turned into an ugly fight. *The Daily Beast*. Available at: http://www.thedailybeast.com/articles/2012/07/23/how-research-into-chronic-fatigue-syndrome-turned-into-an-ugly-fight.html [accessed January 26, 2014].

Science/AAAS, 2014. General information for authors; data and materials availability. *Science Magazine*. Available at: http://www.sciencemag.org/site/feature/contribinfo/prep/gen_info.xhtml#dataavail [accessed October 14, 2014].

Scientific Data, 2014. *Recommended Repositories*. Available at: http://www.nature.com/sdata/data-policies/repositories [accessed October 11, 2014].

Service, C.A., 2014. *CAS Registry*. Available at: http://www.cas.org/content/chemical-substances/faqs [accessed December 27, 2013].

Shobbrook, R.M. and Shobbrook, R.R., 2013. *The Astronomy Thesaurus*. Available at: http://www.mso.anu.edu.au/library/thesaurus/ [accessed December 27, 2013].

Smith, R., 2013. *Elementary Information Security*. Burlington MA: Jones & Bartlett Learning.

Society, A.M., 2014. 2010. *Mathematics Subject Classification*. Available at: http://www.ams.org/mathscinet/msc/msc2010.html [accessed December 28, 2013].

Starr, J. and Gastl, A., 2011. isCitedBy: a metadata scheme for DataCite. *D-Lib Magazine*, 17(1/2). Available at: http://www.dlib.org/dlib/january11/starr/01starr.html [accessed November 2, 2014].

Steen, R.G., Casadevall, A., and Fang, F.C., 2013. Why has the number of scientific retractions increased? *PLOS ONE*, 8(7), p. e68397.

Steinhart, G., 2012. "Lost & found" data. *Flickr*. Available at: https://www.flickr.com/photos/gailst/7824341752/ [accessed September 23, 2014].

Stodden, V., 2010. The scientific method in practice: reproducibility in the computational sciences. *MIT Sloan Research Paper*, 4773–10. Available at: http://papers.ssrn.com/abstract=1550193 [accessed March 21, 2014].

Stodden, V., 2013. What the Reinhart & Rogoff debacle really shows: verifying empirical results needs to be routine. *Victoria's Blog*. Available at: http://blog.stodden.net/2013/04/19/what-the-reinhart-rogoff-debacle-really-shows-verifying-empirical-results-needs-to-be-routine/ [accessed October 14, 2014].

Stodden, V., Guo, P., and Ma, Z., 2013. Toward reproducible computational research: an empirical analysis of data and code policy adoption by journals. D. Zaykin, ed. *PLOS ONE*, 8(6), p. e67111.

Stokes, T., 2013. Mislabeled chemical bottle leads to retraction of liver protection paper. *Retraction Watch*. Available at: http://retractionwatch.com/2013/04/15/mislabeled-chemical-bottle-leads-to-retraction-of-liver-protection-paper/ [accessed January 3, 2014].

Strasser, C., 2012. Thanks in advance for sharing your data. *Data Pub*. Available at: http://datapub.cdlib.org/2012/11/20/thanks-in-advance-for-sharing-your-data/ [accessed January 29, 2014].

Strasser, C., 2013. *Spooky Spreadsheets*. Available at: http://www.slideshare.net/carlystrasser/bren-ucsb-spooky-spreadsheets [accessed April 22, 2014].

Stroebe, W. and Hewstone, M., 2013. Social psychology is primed but not suspect. *Times Higher Education*. Available at: http://www.timeshighereducation.co.uk/features/social-psychology-is-primed-but-not-suspect/2002055.fullarticle [accessed October 11, 2014].

SURF, 2014. *Report. The Legal Status of Raw Data: A Guide for Research Practice*. Available at: https://www.surf.nl/en/knowledge-and-innovation/knowledge-base/2009/report-the-legal-status-of-raw-data-a-guide-for-research-practice.html [accessed October 21, 2014].

Sweeney, L., 2000. Simple demographics often identify people uniquely. *Carnegie Mellon University Data Privacy Working Paper 3*. Available at: http://dataprivacylab.org/projects/identifiability/paper1.pdf [accessed August 12, 2014].

Swoger, B., 2013. Citing data (without tearing your hair out). *Information Culture*. Available at: http://blogs.scientificamerican.com/information-culture/2013/08/23/citing-data-without-tearing-your-hair-out/ [accessed October 31, 2014].

Taylor, B.N. and Thompson, A., 2008. *The International System of Units (SI)*. Available at:

http://physics.nist.gov/Pubs/SP330/sp330.pdf [accessed December 31, 2013].

The J. Paul Getty Trust, 2014. *Getty Thesaurus of Geographic Names*. Available at: http://www.getty.edu/research/tools/vocabularies/tgn/index.html [accessed December 28, 2013].

The Knowledge Network for Biocomplexity, 2014. *Ecological Metadata Language*. Available at: http://knb.ecoinformatics.org/software/eml/ [accessed January 4, 2014].

The National Archives, 2014. *Guidance*. Available at: http://www.nationalarchives.gov.uk/information-management/projects-and-work/guidance.htm [accessed April 25, 2014].

The National Center for Biomedical Ontology, 2013. *NCBO BioPortal*. Available at: http://bioportal.bioontology.org/ [accessed December 27, 2013].

The Open Microscopy Environment, 2014. *OME Model and Formats 2013–06 Documentation*. Available at: http://www.openmicroscopy.org/site/support/ome-model/ [accessed January 4, 2014].

Turi, J., 2014. McMoon's and the lunar orbiter image recovery project. *Engadget*. Available at: http://www.engadget.com/2014/07/19/mcmoons-and-the-lunar-orbiter-image-recovery-project/ [accessed November 17, 2014].

Tyan, D., 2013. How Box.com allowed a complete stranger to delete all my files. *ITworld*. Available at: http://www.itworld.com/it-management/379660/how-boxcom-allowed-complete-stranger-delete-all-my-files [accessed September 23, 2014].

UK Data Archive, 2014. *Storing Your Data*. Available at: http://www.data-archive.ac.uk/create-manage/storage [accessed May 20, 2014].

University of California Curation Center, 2014. *Data Management Plan Tool*. Available at: https://dmptool.org/ [accessed June 24, 2014].

University of Cambridge, 2010. *UK Funding Councils: Data Retention and Access Policies*. Available at: http://www.lib.cam.ac.uk/dataman/resources/Incremental_Cambridge_factsheet_UKfunders_data_policies.pdf [accessed February 2, 2014].

University of Minnesota, 2014. *Data Repository for U of M*. Available at: http://conservancy.umn.edu/handle/11299/166578 [accessed October 26, 2014].

University of Oxford, 2012. *Policy on the Management of Research Data and Records*. Available at: http://www.admin.ox.ac.uk/media/global/wwwadminoxacuk/localsites/researchdatamanagement/documents/Policy_on_the_Management_of_Research_Data_and_Records.pdf [accessed January 28, 2014].

US Department of Energy Human Genome, 2014. Bermuda sequence policies archive. *Human Genome Project Information Archive*. Available at: http://web.ornl.gov/sci/techresources/Human_Genome/research/bermuda.shtml [accessed January 28, 2014].

US Department of Health & Human Services, 2014. *Guidance Regarding Methods for De-identification of Protected Health Information in Accordance with the Health Insurance Portability and Accountability Act (HIPAA) Privacy Rule*. Available at: http://www.hhs.gov/ocr/privacy/hipaa/understanding/coveredentities/De-identification/guidance.html [accessed July 26, 2014].

US Department of Veterans Affairs, 2014. *VA National Formulary*. Available at: http://www.pbm.va.gov/nationalformulary.asp [accessed December 27, 2013].

US Geological Survey, 2014a. *Biocomplexity Thesaurus: Controlled Vocabularies*. Available at:

http://www.usgs.gov/core_science_systems/csas/biocomplexity_thesaurus/controlled_vocabularies.html [accessed December 28, 2013].

US Geological Survey, 2014b. *Biocomplexity Thesaurus: Dictionaries and Glossaries*. Available at: http://www.usgs.gov/core_science_systems/csas/biocomplexity_thesaurus/dictionaries_glossaries.html [accessed December 28, 2013].

US Geological Survey, 2014c. *Biocomplexity Thesaurus: Thesauri*. Available at: http://www.usgs.gov/core_science_systems/csas/biocomplexity_thesaurus/Thesauri.html [accessed December 28, 2013].

US Geological Survey, 2014d. *USGS Biocomplexity Thesaurus*. Available at: http://www.usgs.gov/core_science_systems/csas/biocomplexity_thesaurus/index.html [accessed December 28, 2013].

US Geological Survey, 2014e. *USGS Thesaurus*. Available at: http://www.usgs.gov/science/about/ [accessed December 28, 2013].

UW-Madison Research Data Services, 2014. *Using Spreadsheets*. Available at: http://researchdata.wisc.edu/manage-your-data/spreadsheets/ [accessed April 22, 2014].

Verfaellie, M. and McGwin, J., 2011. The case of Diederik Stapel. *Psychological Science Agenda*. Available at: http://www.apa.org/science/about/psa/2011/12/diederik-stapel.aspx [accessed September 14, 2014].

Vines, T.H., Albert, A.Y.K., Andrew, R.L., Débarre, F., Bock, D.G., Franklin, M.T., Gilbert, K.J., Moore, J., Renaut, S., and Rennison, D.J., 2014. The availability of research data declines rapidly with article age. *Current Biology*, 24(1), pp. 94–97.

W3Schools, 2014a. *JSON Tutorial*. Available at: http://www.w3schools.com/json/ [accessed January 5, 2014].

W3Schools, 2014b. *XML Tutorial*. Available at: http://www.w3schools.com/xml/ [accessed January 5, 2014].

W3Schools, 2014c. *SQL Tutorial*. Available at: http://www.w3schools.com/sql/ [accessed March 18, 2014].

Ward, M., 2014. Cryptolocker victims to get files back for free. *BBC News*. Available at: http://www.bbc.com/news/technology-28661463 [accessed September 16, 2014].

Wei, M., Grupp, L.M., Spada, F.E., and Swanson, S., 2011. Reliably erasing data from flash-based solid state drives. In *Proceedings of the 9th USENIX Conference on File and Storage Technologies*. Available at: https://www.usenix.org/legacy/event/fast11/tech/full_papers/Wei.pdf [accessed July 21, 2014].

Weininger, D., 1988. SMILES, a chemical language and information system. 1. Introduction to methodology and encoding rules. *Journal of Chemical Information and Modeling*, 28(1), pp. 31–36.

Wellcome Trust, 2005. *Guidelines on Good Research Practice*. Available at: http://www.wellcome.ac.uk/About-us/Policy/Policy-and-position-statements/WTD002753.htm [accessed February 2, 2014].

Wellcome Trust, 2010. *Policy on Data Management and Sharing*. Available at: http://www.wellcome.ac.uk/About-us/Policy/Policy-and-position-statements/WTX035043.htm [accessed February 18, 2014].

White, E.P., Baldridge, E., Brym, Z.T., Locey, K.J., McGlinn, D.J., and Supp, S.R., 2013.

Nine simple ways to make it easier to (re)use your data. *Ideas in Ecology and Evolution*, 6(2), pp. 1–10.

White House Office of Management and Budget, 2013. *Uniform Administrative Requirements, Cost Principles, and Audit Requirements for Federal Awards*. Available at: https://www.federalregister.gov/articles/2013/12/26/2013-30465/uniform-administrative-requirements-cost-principles-and-audit-requirements-for-federal-awards [accessed March 15, 2015].

Wilson, G., Aruliah, D.A., Brown, C.T., Chue Hong, N.P., Davis, M., Guy, R.T., Haddock, S.H.D., Huff, K.D., Mitchell, I.M., Plumbley, M.D., Waugh, B., White, E.P., and Wilson, P., 2014. Best practices for scientific computing. *PLOS Biology*, 12(1), p. e1001745.

Woo, K., 2014. Abandon all hope, ye who enter dates in Excel. *Data Pub*. Available at: http://datapub.cdlib.org/2014/04/10/abandon-all-hope-ye-who-enter-dates-in-excel/ [accessed December 2, 2014].

Wood, L., 2009. The lost NASA tapes: restoring lunar images after 40 years in the vault. *Computerworld*. Available at: http://www.computerworld.com/article/2525935/computer-hardware/the-lost-nasa-tapes--restoring-lunar-images-after-40-years-in-the-vault.html [accessed November 17, 2014].

World Health Organization, 2013. *International Classification of Diseases*. Available at: http://www.who.int/classifications/icd/en/ [accessed December 27, 2013].

World Wide Web Consortium, 2014. *Extensible Markup Language (XML)*. Available at: http://www.w3.org/XML/ [accessed January 5, 2014].

Wyllie, S., 2012. Laptop containing university dissertation stolen from car at Foxley Wood, near Dereham. *Eastern Daily Press*. Available at: http://www.edp24.co.uk/news/laptop_containing_university_dissertation_stolen_from_car_at_foxley_wood_near_dereham_1_1373060 [accessed January 26, 2014].

Zivkovic, B., 2013. Good night, moon! Now go away so I can sleep. *Scientific American*. Available at: http://blogs.scientificamerican.com/a-blog-around-the-clock/2013/07/25/good-night-moon-now-go-away-so-i-can-sleep/ [accessed January 26, 2014].

INDEX

CPSIA information can be obtained
at www.ICGtesting.com
Printed in the USA
BVHW071053120820
586015BV00013BA/69